TALES FROM A ROBOTIC WORLD

TALES FROM A ROBOTIC WORLD

How Intelligent Machines Will Shape Our Future

DARIO FLOREANO AND NICOLA NOSENGO

The MIT Press
Cambridge, Massachusetts
London, England

The MIT Press would like to thank the anonymous peer reviewers who provided comments on drafts of this book. The generous work of academic experts is essential for establishing the authority and quality of our publications. We acknowledge with gratitude the contributions of these otherwise uncredited readers.

This book was set in Adobe Garamond Pro by New Best-set Typesetters Ltd. Printed and bound in the United States of America.

Library of Congress Cataloging-in-Publication Data

Names: Floreano, Dario, 1964–author. | Nosengo, Nicola, 1973–author.
Title: Tales from a robotic world : how intelligent machines will shape our future / Dario
    Floreano and Nicola Nosengo.
Description: Cambridge, Massachusetts : The MIT Press, [2022] | Includes bibliographical
    references and index.
Identifiers: LCCN 2021058944 | ISBN 9780262047449 (hardcover)
Subjects: LCSH: Technological forecasting. | Technology—Social aspects. | Robots—Social
    aspects. | Artificial intelligence—Social aspects. | Robots—Fiction.
Classification: LCC T174 .F56 2022 | DDC 601/.12—dc23/eng/20220218
LC record available at https://lccn.loc.gov/2021058944

10   9   8   7   6   5   4   3   2   1

Contents

Introduction

Look around you. Do you see any robots? If you live in a rich country, there's a good chance that your answer is, "Yes—a vacuum cleaner." You may see computers, smartphones, tablets, TV sets, voice-powered assistants, and other pieces of technology in the room where you are reading this book. But it's a safe bet that there's nothing around you now that you call a robot, except for that vacuum cleaner.

And yet they should be everywhere. It has been fifteen years since Bill Gates, in a 2008 cover article in *Scientific American*, predicted that robots would soon enter every home, just like personal computers had done a few years earlier.[1] Since then, countless news stories have told us that friendly, collaborative, humanoid robots would soon work as nurses in our homes and butlers in our bars. We were promised autonomous robots that we could send to places we would rather not visit ourselves, where they would figure out by themselves what to do.

We keep seeing hints of a robot revolution just around the corner. For a few years by now, we've been watching with amazement each new video from Boston Dynamics, the US company that makes humanoid and dog-like robots capable of all sort of stunts: walking in the snow, dancing, doing parkour or gymnastics with impressively "natural" movements. When Elon Musk, in summer 2021, announced plans to build a humanoid robot called Teslabot, he made it sound easy. After all, he noted, Tesla and others' autonomous vehicles are robots on wheels. Though driverless cars are not ready for full deployment yet, their development is accelerating the evolution of sensors and of neural networks and algorithms that collect data to make decisions in real time. Interestingly, Tesla is not the only carmaker making a move toward robotics: in 2021, Hyundai bought Boston Dynamics. The technology is there, Musk argued, to build at least a prototype of a humanoid robot that can move out of your house, go to the grocery store, and pick up the items that you've asked for. If it isn't there yet, Musk seemed to imply, it is because no one has tried hard enough or thrown enough money at the problem.

Musk has a history of succeeding where others repeatedly failed—such as sending a spacecraft to space with reusable rockets or making electric vehicles affordable and cool—and he should be taken seriously. And yes, those Boston Dynamics videos are *very* impressive. But despite what those announcements and demos may suggest, there's still a lot to do before having robots in homes, streets, cities—before your room will be filled with robots ready to do work for you and that you can interact with just as easily as you do with your smartphones.

The reason those robots are not pervasive yet—a reason that Bill Gates overlooked fifteen years ago and others keep overlooking today—is that in order to build robots, we need more than the increasing computing power, component miniaturization, and engineering wizardry that gave us computers, smartphones, and smart wearables. We need a new science. The bad news is: a new science takes a lot of time. The good news is: new science is happening in dozens of laboratories around the world—and that is the story we tell in this book.

The fact that there are few or no robots in your home does not mean that there are no robots at all out there. Modern manufacturing would be unthinkable without robots: more than 3 million industrial robots were operational in factories by the end of 2020, 32 percent of them purchased by car manufacturers.[2] Robots keep finding new applications and markets, from logistics to surveillance, from surgery to farming. Drones monitor plantations and guide harvesting. Fleets of wheeled robots move goods day and night in the large warehouses that power e-commerce, while larger siblings load and unload containers on cargo ships in automated harbors. In 2020, people bought more than 19 million robots for domestic and personal use, although they call them vacuum cleaners or lawn mowers.³ Robots are roving, drilling, and flying on Mars, though they are mostly piloted by humans on Earth.

Most of these robots—the ones that we can buy—are built with the technology of the appliances that they are meant to replace; therefore, they have only a limited understanding of their surroundings and limited decisional autonomy. You cannot tell them or show them what to do. You have to program their moves or, in the best case, select preprogrammed actions from an app. And unlike personal computers, they are not general-purpose machines: they may do one thing very well but cannot easily switch to a slightly different task. Ask them to repeat the same action again and again, and they are great. Ask them to improvise, learn from experience, gain human trust, and they get stuck. In other words, robots are great at a few things, but they suck at all the rest. Worse, they suck at things that we—and indeed much simpler animals—can do effortlessly. These limitations became obvious, for example, when a devastating earthquake and tsunami hit Japan in 2011 and a disaster at the Fukushima nuclear plant ensued. Japan, a robotics superpower if there ever was one, tried sending robots instead of humans to check the site of the nuclear accident, only to discover that they were not up to the task. Not even close.

And yet, think for a moment of what robots could do for us—what problems they could solve, what risks they could take for us, what places they could go—if they only resembled living beings a bit more—for example, if they could understand the meaning or function of what they see, if they could engage with us as other people or pets usually do, or if they could autonomously coordinate with other robots to carry out tasks that a single robot cannot handle, as human and animal societies do. What kind of world could we build with those robots?

In this book, we imagine that world, and we tell what is brewing in labs around the world with the help of brilliant and visionary scientists and engineers who want to make it possible.

Every chapter is built around a fictional story set a few decades into the future, intersected with nonfictional accounts of the research that, here and today, is paving the way for that future. A few words on those futuristic stories. First, we are not novelists, and we do not have the ambition to rival any of the great science-fiction literature on robots and artificial intelligence, a genre mastered by great writers such as Isaac Asimov, Philip K. Dick, and Ian McEwan to name a few. Rather, we use fiction to help our readers see the bigger picture and understand what is at stake with robotics research today. Second, we are not trying to predict the future, but neither do we let our imagination run free as a true novelist would. Each story is visionary enough to get you interested in the science and technology behind it, but grounded enough in the reality of current science and engineering to think that, if all goes well and all of roboticists' efforts pay off as they hope, it would be technically possible for that scenery to take place in that specific time frame. Third, just like the nonfictional parts of the book, the fictional stories owe a lot to the work of the international robotics community, building on ideas and visions that are shared at conferences, in publications, and in informal discussions with colleagues.

The kind of robotics that this book is about has a bright future but a relatively short history, which is worth summing up before we begin telling our tales.

The history of robotics can be roughly divided into three phases. Initially, in the 1960s and 1970s, robots were conceived as machines capable of automating repetitive human work in labor-intensive industries. Industrial robots were, and still are, prized for high precision, relentlessness, power, and speed. A decade after World War II, industrial robots enabled the production of goods of better quality at lower prices on a larger scale and at the same time relieved humans from strenuous and dangerous jobs. The automation capabilities of industrial robots kept improving over the past sixty years and found applications in a variety of industries that produce, refine, assemble, and package goods at an intensive pace. According to the International Federation of Robotics, installations of industrial robots grew by 13 percent in 2020, hitting a record of 3 million robots operating in factories around the world.[4] Most of these robots are so fast and powerful that they must work in caged areas to prevent the risk of harming humans. Although many have sensors for adjusting their moves (for example, to precisely position a screwdriving head), they are not designed to cope with unforeseen situations and make autonomous decisions.

Early attempts in the 1970s to extend the control methods of industrial robots to autonomous locomotion in unrestricted environments resulted in slow and ungainly movements that would work only in very simple and controlled environments. Their brains were busy building and continuously updating mathematical models of their surroundings in order to plan the next actions, and those calculations were possible only for simple environments and moves. Those robots were bulky and had limited computational power, and crafting their artificial intelligence required sophisticated mathematical and engineering skills.

The situation changed some twenty years later, in the early 1990s, when the nascent industry of personal computers and accessories democratized programming languages and led to the commercialization of small electronic components, sensors, motors, and computing chips at affordable prices. A second robotics phase began, and a new generation of programmable small robots, mainly wheeled robots, started to populate departments of computer science and mechanical engineering at universities around the world. At that time, some computer scientists challenged the applicability of traditional control methods for robots that must operate in unforeseen and changing environments. Joined by cognitive scientists, neuroscientists, ethologists, and philosophers, these researchers claimed that intelligent behavior results from the parallel activation of several simple, stimulus-response actions triggered by the interaction between the robot and the environment. According to these scientists, animals and robots required only a suitable set of sensory-motor responses, such as "turn left if obstacle detected on the right," "if bright light detected, steer in the corresponding direction," "if no stimulation for ten seconds, perform random motion," instead of building and continuously updating sophisticated mathematical model and plans,

Rodney Brooks, then a professor at MIT and one of the most prominent advocates of this novel approach to artificial intelligence, proposed a "behavior-based" programming language whereby robots can be continuously updated with new hardware and behaviors as needed. Other researchers resorted to artificial neural networks, which at that time had started to boom, to connect sensors and motors and let the robots learn and evolve as biological systems. This was a period of intense intellectual activity that still drives much of today's robot research. The mechanical design and intelligence of those robots were often inspired by nature, and biologists started to use robots as behavioral models of living systems. The biological inspiration was most often taken from insects and relatively simple vertebrates whose behaviors could be loosely replicated. Those robots were prized for the capability of operating in unforeseen and changing environments, for adaptability, fast reactions, and behavioral autonomy rather than predictability. This approach led to the commercialization of wheeled robots that clean houses, mow the lawn, transport food in hospitals, and patrol industrial plants; of drones that let us take stunning aerial photographs, monitor crop fields, and inspect industrial infrastructure; and of underwater robots that survey ocean pollution and help preserve marine life.

The sale of industrial, personal, and service robots has been growing ever since, and according to most economic projections, they will continue to grow at a double-digit annual rate. Watching a robot clean your house or a drone follow you while you jog, it is difficult to abstain from attributing to them some sort of intelligence. But they still look and feel like versions of appliances that we are familiar with: vacuum cleaners and remote-controlled aircraft. Today's robots affect our life, for good or bad, but our life does not depend on them; they replace us at dirty, dangerous, and boring jobs, but we do not shed a tear if they break down; they do their job day and night in our homes and workplaces, but we do not engage with them as we do with our pets, friends, or colleagues. They buzz around the floors of e-commerce warehouses in thousands, but we do not expect them to build a home where they find refuge and have social interactions as animal and human societies do.

Almost twenty years after the seeds of those autonomous robots were sowed, we are now entering a third phase, where a new generation of intelligent robots is emerging from research labs and start-ups that leverage a unique combination of recent scientific and technological developments. Brain-like learning algorithms now allow machines to learn skills that match and sometimes surpass human abilities, and their development is massively supported by an unprecedented investment of governments and companies. Computing chips with neural architectures recognize faces and places at electronic speed. Soon, 5G wireless technologies will allow swarms of robots to communicate with each other almost instantaneously. Plastic and metal parts, which today's robot bodies share with dishwashers and cars, are replaced by soft materials that do not merely serve as scaffolds, but also integrate sensing and actuation, akin to biological tissues endowed with sensors, nerves, and muscles. Biodegradable and edible components, neural implants, and metabolic energy production redefine the boundaries between the living and the artificial. This new wave of soft robots can no longer leverage classic control theory, whose mathematical models are computationally tractable only for rigid bodies connected by rigid joints; rather, they are a fertile ground for neural networks that learn how to make sense of the richly sensorized and innervated bodies that flex, twist, roll, fold, and curl like living organisms. Neuroscientists and bioengineers are starting to understand how to communicate with our nervous system, paving the way to a new generation of brain-interfaced robots that can assist, replace, or augment humans as if they were a natural extension of our bodies.

This book is about the novel generation of robots that is emerging from labs around the world. Interestingly, new waves of robots do not replace previous ones; rather, they coexist, finding new application areas and interbreeding to improve, as we will find out in some of the stories.

One last note on how we look at the future. There is much hype surrounding robotics and artificial intelligence or, for that matter, nanotechnology, neuroscience, quantum computing, all fast-paced and innovative areas of science. In our fictional scenarios, we focus on how robots can help us tackle some of the biggest challenges that humanity will have to face over the next few decades: global warming, natural and man-made disasters, an ageing society, the search for new habitats beyond our planet, love and fear of intelligent machines, and abuse of natural resources, to mention a few. In our stories, we mostly look at the bright side and show how robots could solve problems. We are not so naive as to suggest that technology by itself can save the world. All of those challenges are complex problems that will require political, economic, and societal measures alongside technological ones. Robotics cannot be the silver bullet, because there will be no silver bullet for any of those problems. Nor are we that naive to not see that along with new opportunities, radical innovation in robotics and artificial intelligence will also create new problems and new societal challenges. In the final pages of this book, we ask how future robotics may affect welfare, employment, and the distribution of power in society. In one chapter we look into the possible impact of robotics on jobs, wealth, and inequality. In another one we explore what structure the robotics business and market may take in the future, and in the last chapter we consider how things could go wrong—inside and outside laboratories—and tell what researchers are doing to prevent that.

A final disclaimer. Every year, more than five thousand peer-reviewed articles appear in major conferences and journals dedicated to robotics. The research highlights and scientists depicted in this book are just a small sample—with some inevitable bias towards our closest network of colleagues—that, for sake of space and time, we subjectively selected to capture the broader movement of ideas and technologies that we believe will define the new generation of robots. We hope that readers will be intrigued and decide to explore and contribute to the definition of tomorrow's intelligent machines.

SUGGESTED READINGS

Brooks, Rodney. *Flesh and Machines*. New York: Pantheon Books, 2002.
> A highly readable introduction to biologically inspired robots and their interactions with humans.

Darling, Kate. *The New Breed*. New York: Holt, 2021.
> An intriguing analogy of our relationship with animals and with the new wave of intelligent robots.

Floreano, Dario, and Claudio Mattiussi. *Bio-Inspired Artificial Intelligence*. Cambridge, MA: MIT Press, 2008.
> An introduction to methods for instilling biologically inspired intelligence in computers and robots.

Kelly, Kevin. *Out of Control*. Boston: Addison-Wesley, 1994.
> A thought-provoking book on the new wave of bioinspired artificial intelligence and machines.

Pfeifer, Rolf, and Josh Bongard. *How the Body Shapes the Way We Think*. Cambridge, MA: MIT Press, 2006.

On the importance of embodiment in biological and artificial intelligence.

Sejnowski, Terry. *The Deep Learning Revolution*. Cambridge, MA: MIT Press, 2018.

An accessible introduction to the new wave of artificial neural networks and their applications.

Siciliano, Bruno, and Oussama Khatib. *Handbook of Robotics*. Berlin: Springer, 2016.

Systematic coverage of all fields of robotics written by renowned experts in accessible prose for both researchers and aspiring roboticists.

Winfield, Alan. *Introduction to Robotics*. Oxford: Oxford University Press, 2012.

A concise and factual description of what robots are and can do.

1 ROBOTS IN THE LAGOON

As he looks out of his window overseeing the Giudecca Canal, Enrico cannot help but think of that day, many years ago, when his parents told him that they were all going to leave Venice for good. He could not stop crying. He was only seven years old, and the city's alleys and canals were his whole world. But his parents had made up their minds—and thinking back to how things looked at the time, who could blame them? Sure, over the centuries Venetians had gotten used to *acqua alta*, the high tides that occasionally flooded the lowest parts of the city, including Piazza San Marco. But by the early 2030s, such tides had become so frequent that Venetians were spending a good part of the autumn draining water from the ground floors of their houses and shops and assessing the damage. In few other places in Europe were the effects of global warming so visible and so disrupting.[1] Rising sea levels and increasing precipitation were conjuring to flood the city up to twelve times a year, with water levels reaching 140 centimeters (the threshold for "exceptional tides" back in the twentieth century) on average once a year. The system of mobile gates anchored to the seabed designed in the distant 1980s to protect the city was only a partial solution and was proving no match for twenty-first-century sea levels.[2] When the historic high tide of 2033 hit, thousands of Venetians decided that they had had enough. Venice seemed poised to become a ghost town: crowded by tourists during the spring and summer but empty of residents apart from hotel managers and gondoliers.

Enrico's mother and father had started looking for a house on the mainland, but their child's endless crying and foot stomping at the idea of leaving Venice was opening some cracks in their resolve. In autumn 2035, when the local administration and the Italian government announced an extraordinary plan to do "whatever it took" to save the city and asked residents to believe in it, Enrico's family was among the ones who chose to believe—and to stay.

"The best decision my parents ever took," Enrico now thinks as a tide that will probably end up in history books materializes before his eyes. Heavy rain has kept falling for three consecutive days. Swollen rivers are washing into the lagoon, while a strong Scirocco—a warm wind from the southeast that often blows in the autumn in Italy—keeps pushing more and more water from the sea and toward the city. It is the same combination of events that caused the 1966 flood he has read about so many times, as well as the 2033 one he remembers firsthand. And yet like all his fellow citizens, he is confident that waters will not flood Piazza San Marco this time and will not invade houses and shops.

What will hold back this tide and has in fact changed the city's whole destiny is an unprecedented and ambitious application of swarm robotics, a field that started moving its first steps in the early 2000s and would later revolutionize the way robotic systems were built.

When the project was first described in the 2030s, many Venetians were skeptical—to say the least. Of all the crazy ideas that had arrived in response to the city's call for proposals for new systems that would protect Venice from tides, that was by far the craziest one. It envisioned hundreds of robots swimming across the lagoon like fish schools, constantly gathering data on water circulation and quality, checking the submerged foundations of historic buildings and fixing them when necessary, and, most important, building dams on the spot when tides hit. For sure, it was a big change from the previous defense, the MOSE system of retractable gates anchored to the seabed conceived back in 1981. Originally scheduled to become operational in 2011, it was delayed until 2022, and by that time many of the gates were already plagued by corrosion. Its seventy-eight gates somehow protected the lagoon from extremely high tides, but were not of much use against medium tides, those between 80 and 100 centimeters, which were far more common:

it could not be deployed quickly enough to stop them, and using it so often would have been too expensive anyway, and not only in economic terms: researchers checking on the lagoon bed as early as the late 2010s had found a worrisome erosion caused by the project. MOSE was not a bad idea in itself, but it was a twentieth-century idea: a large, expensive technology with a heavy impact on the environment, designed without knowing how global warming was changing the rules of the game.

What a contrast with what roboticists were proposing to the city council: a modular, reconfigurable system, made of small robotic bricks that would cling to the seabed only for the time necessary to stop a tide—even a medium-sized one—and then leave it alone, moonlighting as sentinels of the city's health for the rest of the time. Adaptable, light, and relatively inexpensive, it was the kind of solution the twenty-first century required.

Sure, it was a crazy idea, but a captivating one for Venetians, who in the end chose to bet on it. When it came to convincing them, it helped that by that time, their city had already been an open-air robotics laboratory since the late 2010s, when an Austrian zoologist had the idea of creating a robotic ecosystem in the muddy waters of the lagoon, let it evolve, and see what would happen.

GRAZ, AUSTRIA, CURRENT DAY

Compared to your average zoologist, Thomas Schmickl spends surprisingly little time in the company of real animals. A quick look at his curriculum vitae suggests that since the 1990s, when he was working as a software developer while studying to graduate in zoology at the University of Graz, the Austrian city where he was born and raised and where he now heads the Artificial Life Laboratory, he has often had problems deciding whether he was more interested in the natural world or the artificial one.

Even after choosing zoology as his main trade, Schmickl realized that his programming skills could still prove handy. His main interest as a biologist lies in understanding how collective behavior is organized in animals: how members of the same species can exchange information, influence each other's behavior, and ultimately work toward a common goal—be it finding food, building a nest, or flying in formation for thousands of miles. But how do you answer such questions? For zoologists, devising ways to reliably test scientific hypotheses on animals' group behavior in the wild is typically a headache. A scientific test requires controlled conditions—but how do you fly a flock of birds in controlled enough conditions so that you can work out who among them decides when to change direction and how this coordination works across all swarm members?

One way to do it is by using computer algorithms to create a simulated flock: you give "virtual birds" rules based on what you think actual animals are doing, let the program run, and see if the results resemble what you observe in a real flock. A widely used computational model of swarm behavior introduced in the mid-1990s by the Hungarian physicist Tamás Vicsek, for example, has been used to account for how starlings fly in formation, suggesting that what guides each bird's flight is the constant search for a spot inside the flock where the density still allows it to see enough of the surroundings instead of being blinded by other birds.[3]

Things become much more interesting, though, when you replace virtual agents with actual robots modeled on the animals you are interested in. For a biologist, this means that group behavior can be simulated in a more realistic way, factoring in how the environment and physical interactions between the animals influence their behavior. For roboticists, the reward is even bigger: a group of dozens, even hundreds, of simple robots equipped with the same ability to self-coordinate that we see in birds and bees would be more powerful and more robust that any individual robot could ever be. Such a robotic swarm—or flock, or herd–could monitor a larger area more quickly and better than a single robot; its members could pass information to each other to constantly improve their performance, and the system as a whole would be more robust; just like the Internet survives if some of its nodes stop working, a robot swarm can keep functioning when a few robots fail.

That is why Schmickl, like many others in the 2000s, became interested in swarm robotics: the scientific attempt to recreate with robots the collective behavior observed in social animals. The concept of swarm intelligence (artificial systems inspired by the collective behavior of animals) had been around since the 1980s, and by the mid-2010s, swarm robotics had been a field of intense research. "But there were still almost no robot swarms deployed in the real world—only in labs or in controlled conditions," Schmickl explains. "If we cannot take our robot swarms outside the lab, we will never be able to prove their worth. That is why I wanted to show that you can go out of the lab and have a swarm of robots interact in a living city, and show real applications for them."

The Venice lagoon offered a unique opportunity for that: in almost any other city in the world, unleashing dozens of robotic units on the streets would disrupt traffic, scare people, or result in robots being stolen and damaged. Or, most likely, regulatory hurdles would block the project before reaching that stage. But Schmickl reasoned that in a city where the streets are canals, filled with relatively calm waters, robots could swim freely and hidden from sight, undisturbed by citizens and without disturbing them. It would become possible to put together a robot swarm of unprecedented size and let it change and evolve over time, possibly mimicking, at least in part, the complex dynamics of a real ecosystem. "There are many aspects that brought us to Venice," Schmickl explains. "The water is calm but murky, which creates challenges for robot sensing. The channels are a sort of maze, which is interesting for cognitive experiments. The whole lagoon is quite large, which means you have to cover large distances, but it is not open water: your robots cannot get lost easily." In addition, robots can carry sensors, and few cities are more in need than Venice of having their health closely monitored. Its lagoon is a complex, ever changing environment, home to diverse species—from algae to molluscs, from fish to birds—in constant interplay with human activities. And on it or, rather, in it lies a treasured city that risks being drowned by the same waters that make it unique.

Schmickl teamed with researchers from Belgium, Italy, France, Germany, and Croatia, and in 2014, they won funding from the European Commission to start the subCULTron project, an ambitious plan to put 120 custom-built robots in the lagoon waters . The team built three families of robots, each one inspired by a real, living inhabitant of the lagoon. In the first group are fixed, mussel-like robots that stay on the sea floor and can monitor the presence and passage of algae, bacterial colonies, and fish. They can sense a lot but can move only a little—basically only to resurface when needed or to dock to the sea floor. The second group, floating on the water surface, is made of robotic lily pads that gather energy from sun rays, receive information from satellites, and record water movements to measure tides and detect the passage of ships. In between, underwater swimming robots (equivalent to fish in the real ecosystem) continuously explore the lagoon environment and exchange information with both mussels and lily pads, downloading data from the former and docking to the latter to upload it and recharge when necessary. Because radio waves—the means of communication of choice for flying and walking robots—do not travel well underwater, the robot fish have to relay information from the mussels to the lily pads, which can then transmit them to the researchers via radio links.

Schmickl's and his colleagues' idea is not only to record data on the lagoon's health but also, in the long run, to let this hybrid robotic society evolve and differentiate. From the exchange of information between different "species" and their interaction with the complex and changing environment of the lagoon, robotic subcultures would at some point emerge. For example, the population around the Giudecca, an island, could learn to exploit currents differently from the population in smaller canals, which in turn would have to negotiate with stagnating waters and the continuous passage of gondolas. The mussels in the tiniest canals would teach their companion fish to stay away from the algae-filled canal bottoms, a problem unknown to the outer areas of the lagoon.

Collectively, the robots could end up knowing the lagoon better than Venetians themselves and help them figure out what to do to save it. The subCULTron project is a first step toward the creation of a "robotic society" in the lagoon, which Schmickl hopes will convince skeptics of the real-world potential of swarm robotics.[4] Also of its other-world potential, in fact, is a possible future evolution of the project to use similar robot swarms to explore distant moons and planets such as Io, Europa, or Titan. "It is foolish to send single robots up there, putting all your eggs in one basket," he says. "We should be using robot swarms." But before that, he hopes his robots or their descendants can contribute to help preserve the very city that is hosting them—a city that faces an uncertain future, as previous plans to protect it,

such as the controversial MOSE system, seem to be failing. "We thought a lot about that, and I hope robotics can offer solutions in the future."

VENICE, ITALY, 2051

The rain does not stop, and the sky keeps getting darker and darker so that Enrico can no longer see from his window what is happening out there in the lagoon. He turns on his smart glasses and goes to the city's data cloud. For most Venetians, a visit to the lagoon section of the cloud is part of a daily routine. Gathering and integrating data from the hundreds of robots floating and swimming through the lagoon, it shows what is going on in every canal: how clean the water is, where algae are building up, where the foundations of a building are in need of repair. But it is during tides that the data cloud gets most of the traffic.

An interactive map is showing the tide situation in every canal, the forecast of how water will rise in the next forty-eight hours, and previewing what the robots are going to do to stop it.

Over decades, thanks to Thomas Schmickl's pioneering work and subsequent follow-ups to his project, more and more robots have inhabited the lagoon, getting to know the currents, developing the ability to plan routes through the maze of canals, and collectively figuring out the best place to build a dam, compromising between sea depth, strength of the current, effectiveness in protecting the city, and avoiding damages to the ecosystem down there. As Schmickl and other researchers had predicted, the possibility of gathering data from a real and complex environment such as the lagoon, where natural and human activity constantly overlap, had proved to be a boon for the development of swarm robotics, now extensively applied in manufacturing, agriculture, search and rescue, and even space exploration. But what happens in Venice at every high tide remains one of its most impressive applications.

The map shows hundreds of dots coming out of underwater warehouses, placed at strategic spots where the dam-building robots are stored, and moving toward Bocca di Porto del Lido, one of the three inlets that put the lagoon in communication with the Adriatic Sea and the closest to the historic part of the city. In the real world, each of those dots is a dam-building robot, a sort of robotic sea urchin the size of a basketball that unfolds from a flat disk within the warehouse into a pulsating ball, regularly shrinking and expanding its artificial skin populated by tiny tentacles that make it gently roll on the lagoon floor. All robots move toward the same lagoon inlet, attracted by a low-frequency sound picked up by the sensorized tentacles. That sound, the emergency call emitted by sentinel robots submerged at Bocca di Porto del Lido, operates at a frequency that can travel several kilometers through the murky lagoon waters without disturbing the aquatic biological fauna. When the robotic sea urchins arrive by the inlet, the tiny magnets at the extremities of their tentacles lock into other robots as they continue to rhythmically expand and contract, albeit at a slower pace. More robotic sea urchins arrive, cling to the robotic cluster, and roll upward when the cluster contracts, while those below stay in place securely anchored to other robots through their tentacles. Within a couple of hours, a slowly pulsating structure, like an artificial lung, starts emerging from the waters across Bocca di Porto del Lido, creating an effective barrier against the rising waters. Once the tide is over, the sentinel robots will change the sound pattern to signal that the danger is over, and the pulsating sea urchins will retract the magnetic spines that hold them together. The dam will be gradually dissolved by the waves, and the robots will gently roll back to the warehouse guided by sound. It is a wonder of self-organization that took scientists several decades to conjure up.

The biggest technological breakthroughs often come when different innovations cross paths and start working together. All of a sudden, things that seemed distant and related begin to appear as if they were made to work together. It had happened at the beginning of the twenty-first century, when mobile telephony, touch screens, digital cameras, accelerometers, and a bunch of other seemingly unrelated technology were mixed and stirred into a single cocktail by Apple founder Steve Jobs and became the iPhone. And it happened again in the late 2020s, when a whole new generation of robotic technologies was finally ready to get out of the labs and creative entrepreneurs began mixing them in new ways, turning them into the building blocks of the robotics revolution that took the 2030s by storm. A typical example was how swarms of sensing and "culturally evolving" robots, such as those pioneered by Schmickl in Venice, crossed path with modular, self-assembling robots that other researchers had been testing more or less at the same time in the United States.

NEW YORK, NEW YORK, USA, CURRENT DAY

Hod Lipson's path to robotics is somehow complementary to Schmickl's: whereas Schmickl is a biologist who has chosen robots as the tools of his trade, Lipson is an engineer who uses machines to get to the heart of how life works. Born in Haifa, Israel, where he graduated in mechanical engineering from the famed Technion Israel Institute of Technology, Lipson has lived and worked in the United States since the early 2000s, first at Cornell University and then at Columbia University, which he joined in 2014. Over the years, his research interests have spanned fields as diverse as 3D printing, neural networks, driverless cars, and various styles of robotics. But when it comes to robotics, he says, "My real interest has always been in imitating life, not so much in making a robot that's useful. It's a quest to recreate life with artificial means, very much like alchemists wanted to do a thousand years ago." Recently he's been obsessed (an obsession that was widely recounted in magazine articles, video interviews, and TED talks) with creating self-aware robots. In what is now probably his most famous project, he coupled a simple robotic arm to a neural network whose task was to create a computer simulation of the arm itself.[5] Roboticists routinely use simulations of their robots to test their control algorithms, and it is common to infuse autonomous or semiautonomous robots with a software model of themselves and their environment to help them navigate and make choices. But such models are typically built using good old-fashioned computer programming. "We have been using neural networks for some time to let robots learn about their environment, but oddly enough, we have not used them to help robots learn about themselves." This is what Lipson did, using deep learning as a tool for robotic introspection of some sort. Like most other deep-learning-based artificial intelligence (AI) systems, his network would start with almost random attempts, creating a digital representation of something that had no resemblance to the actual robotic arm. But over thousands of learning cycles and using the measurements of the actual movements and positions of the arm to self-correct, it ended up with a quite faithful digital replica of the robot. Lipson suspects this process has a lot in common with what happens in a child's brain during early development. Robotic self-awareness, he thinks, could shed light on how consciousness arose in humans and could help build machines that can more easily adapt to changing conditions—though in interviews about the project, he regularly ends up having to fend off fears of a conscious, Terminator-like, superintelligent robot.

At the other end of the spectrum, though, Lipson is equally interested in machines with very little—if any—intelligence, which can do wonders when combined in large numbers. In 2019, in collaboration with Daniela Rus at MIT, he published a paper in *Nature* magazine describing a "particle robot," loosely inspired by the way individual cells in our body team up.[6] Each particle in this robot is a small disc that can only do one thing: rhythmically expand and contract. A tiny magnetic sphere lined up along its outer ring allows each disc to bind, but never too strongly, to other discs, and it is then that things become interesting. Whereas a single particle cannot move from its location, as soon as many of them attach to one another, the resulting robot can indeed move around. To control this movement, Lipson and his team mounted a light sensor on each disk, turned on a single light source, and programmed the particles so that the offset of the respective phases of contraction and expansion would be proportional to the intensity of the light received. In practice, if the particles that are closest to the light source start pulsating at a certain moment, the ones that are a bit farther will start a few moments later, those that are even farther will have a yet larger delay, and so on. The result of this interplay between light-trigger motion and delayed lighting is a surprising collective movement of the molecular machine. The whole robot can move toward a light source, transport objects, and even find a way around obstacles. And yet on its own, no particle can go anywhere, and there is no communication among them.

"We call them *particles* precisely to emphasize that the individual elements have no intelligence," Lipson explains, "like molecules in our bodies, and also to highlight the statistical nature of the system. Molecules and cells in our bodies are not moving in a precise and prescribed way; they are mostly knocked around. And yet despite this, we see pretty robust systems that behave mostly in a deterministic way and can self-repair." The simplicity of the particles and the fact that they do not even need to communicate with one another makes the system particularly robust and scalable. At the end of the day, it does not make a lot of difference if the robot is made out of ten particles (as in Lipson's real experiment) or ten thousand (as in the computer simulations he did for the same paper).

Lipson insists that he has no practical application in mind for now. "We are sailing West, as we say in the United States," he quips. "I guess there is something interesting on the other side, and I am starting to explore."

Still, he suggests that in that robotic West, a new way to design and build machines could be hiding. "There is a growing realization among roboticists that we have been building robots the wrong way. Now they are made with a monolithic, top-down approach. If we can build them bottom-up, with small pieces, they will be more robust, more easily adapted to various applications. Here, this idea is taken to an extreme, where we don't even need that tight integration between modules, or the communication between them that is typical in other modular robotics approaches. I think if we keep pushing this metaphor to its extreme, we can get to interesting places."

The next steps, according to Lipson, would be to make the particle tridimensional and a lot smaller. In our fictional Venice in this chapter, we've imagined these particles growing bigger, to the point where they could make a dam. But Lipson would be equally interested in making them as small as a millimeter or less, like a grain of sand. "If we can make a million of them, we would really see interesting things." That would mean changing materials and manufacturing methods, using chemical or acoustic rather than mechanical actuation. It would also be interesting to add more sensors and computation on each particle. "It's entirely feasible. You can go at the millimeter scale and still put a web server on each particle."

In the end, Lipson views the shift toward bioinspired mechanics as a natural companion to artificial intelligence and as a necessity toward a sustainable future. "Neural networks, which are the state-of-the-art in AI today, are bioinspired. For a long time, it was not obvious that they would be successful, and many other approaches were tried that were not bioinspired, like expert systems". In the end, Lipson notes, those approaches failed and neural networks succeeded, and there is a lesson for the physical part of robots there too. There is a temptation to stick to electric motors and titanium parts, because roboticists already know how to build them. "But in the end, we'll have to go bioinspired, and that means modular. It's the only way you can get to a robot that builds another robot, and in order to have a sustainable robotic ecosystem, you need robots that build and fix other robots, and recycle and incorporate parts of older robots into themselves. Today you cannot recycle much of a robot, maybe some aluminum parts. But if robots are made of small, very simple components, they can be recycled into another robot. This is how biology works."

VENICE, ITALY, 2051

Less than four hours after the acqua alta alarm first rang out across the canals, the robot builders have finished their job. The dam is now visible from Enrico's window, a thin black line that stretches along the Lido inlet, effectively sealing it and holding back the water.

The robot swarm itself will decide how long it will stay there, and whether at some point it will need backup to reinforce the dam—although the final decision will be in the hands of the city's authorities. The city has several thousand dam robotic elements, enough to face very high tides. Should tide levels get to 200 centimeters, surpassing even the historic 1966 flood, it will still be possible to borrow more from Rotterdam. Five years ago, the Dutch city adopted a twin system to protect itself from the waters of the North Sea. The success of the Venice robots has not gone unnoticed: from Lagos to Hong Kong, from Melbourne to Houston, other cities threatened by rising sea levels are now experimenting with similar systems.

Enrico has had enough of watching all this from his window. Just like when he was a child, there is nothing he loves more than walking the alleys and bridges of his city—and rain or not, he will do it today too. He fetches his hat and umbrella—but not his high boots, as his parents would have had to do in such a day—and gets out.

Despite the rain, the alleys around Piazza San Marco are as alive, crowded, and noisy as any sunny day. Apart from gondoliers, who will have to wait for the end of the storm to resume working, citizens and tourists go on with their lives unconcerned. A tide that a few years back would have flooded half of Venice and blocked the city for days has been kept off by robots—descendants of those first settlers that entered the lagoon in the distant 2010s. And a city that was on its way to becoming a casualty of climate change has turned into a poster child of how the right mix of technologies can mitigate its effects.

2 THE REALLY BIG ONE

The event depicted in this chapter is, by definition, unpredictable. Scientists say that the kind of earthquake we imagine here has a significant probability, ranging from 10 to 37 percent, depending on the estimate, of happening within the next fifty years.[1] That means it could happen any day within that time frame, or it could happen much later, maybe a hundred or two hundred years from now. But if seismologists are right, in the long run, the probability becomes 100 percent. It will happen at some point. We have set this chapter in a relatively close time to highlight that the technology we describe—technology that could indeed make some difference in such a catastrophe—is not so far away. We strongly hope this disaster will not happen so soon, but if it does, the technology could be there. It will take political will, more than major scientific breakthroughs, to make sure it is ready when it is needed.

9:10 a.m., Seattle Waterfront

"Can we tell the waitress to hurry up?" she said, a bit nervously. "I'll end up running late."

"Relax," he replied, placing his hand on hers, and nodding with his head to Elliott Bay and the ocean and the bright blue sky in front of them. "How many days like this do you think we'll have before autumn begins? This is a gift. The world will not stop turning if you get to the office a bit later, will it?"

As far as weather was concerned, he was right. It was an exceptionally warm day for late September in Washington State. That morning, after getting up and looking out the window, Michael had managed to convince Liz to have breakfast here, in front of the sea, at the same bar on the Seattle waterfront where they often spent Sunday mornings in the summer. They would both be a bit late for work, but what the hell? As a software engineer at one of the two tech companies that basically owned the city, he could make his own schedule—though in the end, he basically lived in the office. As emergency operations manager with the Seattle office of the Federal Emergency Management Agency (FEMA), she had less flexibility and worked in shifts. But hers would only start at ten that day—though she always tried to arrive at the office half an hour earlier so that her colleagues could fill her in on whatever was going on.

"Okay, you are right. Let's take this gift" she conceded, taking a deep breath in the cool ocean breeze just when the waitress arrived and delivered two cappuccinos at their table.

The summer that was just ending had been a happy time for them. Their relationship had just really taken off after beginning earlier that year, and in June she'd moved in with him. Their careers were also in full swing. Earlier that spring, she had been put in charge of overseeing the emergency plans for the seismic risk division. "How does it feel to work all the time to prepare for something that may never happen?" he once asked her. "Well, it will happen at some point," she replied. "Maybe not in our lifetime, but there will still be humans around here when it happens. The plans I make will save some lives. Maybe our children's lives."

They'd started talking about having a child in fact, half-jokingly at first, then more and more seriously. They felt ready for it, though it would mean that they would both have to stop working such long hours. He had also gotten a promotion in May, becoming deputy head of the cloud computing division, and he was now overseeing the infrastructure that powered much of the western states' business.

"What will your day be like?" she asked him, as they sipped their cappuccinos.

"A lot of chatting and patting on backs, followed by faking interest in keynote presentations I guess," he laughed. "There's a big meeting later this morning. The marketing guys will present their research on what our cloud customers want, the R&D guys will showcase their latest ideas on new things our customers may want but don't know yet, and in the end the big bosses will choose two or three new products. And you?"

"The usual, minus the patting on backs. Routine checks on all our monitoring stations. A teleconference with our colleagues down in California. They've just made some updates to their emergency plans, and we want to check if there's something interesting for us too. And I hope to make some time to read a paper that just came out on the 2011 earthquake and tsunami in Japan. We're still learning a lot about it, and it's the closest to what we can expect here."

She looked at the watch, pretending to be relaxed.

"Okay, I can't hold you anymore" he smiled, waving at the waitress. "I'll take the check. Just go. I'll see you at home tonight."

"Thanks. It was a great way to start the day. We should do it more often." She kissed him, got on her motorbike, and rode away—a bit too fast, for his taste—toward the FEMA offices, located about half an hour away in Bothell, a small town in the outskirts of Seattle. He paid the bill, stayed a bit longer. and walked to the company headquarters, only a ten-minute walk from the waterfront.

10:53 a.m., Nile Headquarters

It was well before 11:00 a.m., but Michael was already in the large meeting room on the seventh floor. He was there first, wanting to make sure that the cool videos he had prepared for his presentation would work on the sometimes unpredictable projector in that room. And also—as he jovially told Steve, the marketing director from New York City, when he entered the room a few minutes later—to enjoy the view of the harbor from the large panoramic window, which would have to be obscured once the projector was turned on.

He and Steve were good friends but had not seen each other for almost a year and had a lot of catching up to do while waiting for the others to arrive. They poured themselves some coffee and started chatting on who had been promoted and who had not, who had moved to the Shanghai office and who was coming back to the United States, who was expecting a child and who was getting divorced. Michael was just about to tell Steve that he and Liz were getting serious, when he heard a clinking sound from the table where glasses, coffee, and fresh juice had been lined up.

He turned around, wondering if he'd inadvertently bumped into the table. One second later, the table bumped into him and the whole room began to shake. Furniture was moving around like it had wheels. The framed photos on the wall fell, their glass covers shattering and mixing with the drinking glasses and the LCD screens, all thrown violently on the floor. Michael, with the presence to remember his emergency training, grabbed Steve by an arm and they both ducked under the large meeting table at the center of the room. After a seemingly never-ending three minutes, the building kept oscillating for a while and gradually came to a halt. Michael looked at Steve, whose face had literally turned white, and asked him, "Are you all right"? Steve forced himself to answer, unconvincingly, "Yes," but was obviously frozen with fear. Unlike in Seattle, the New York City staff had never received intensive earthquake training in preparation for this day.

Michael realized he would have to lead his colleagues out of there. He grabbed his phone, hoping to get in touch with Liz, but there was no signal. Electricity was also down. Many times over dinner, she had made him rehearse what he would have to do if this moment ever came. He told Steve to stay calm and follow him to the stairs. "I don't think it's a good idea to go out in the street," Steve replied.

"The building is antiseismic, isn't it?" asked Steve. "If it resisted this shock, it would resist the aftershocks too. It can't get any worse." Michael looked at him and tried to sound as calm as he could. "You are right. We do not want to be out in the street. In fact we want to go up, trying to get to the terrace on the top floor. But you are wrong. It can get much worse."

10:55 a.m., FEMA Offices

Liz had spent the last six years of her life, since taking up that job, imagining how that moment would play out, rehearsing what she would have to do.

And yet for the first few moments after the building had stopped shaking, after the seismometers stopped drawing waveforms that reached the limits of their scales, after everyone had reemerged from under their desks, she remained silent and still for a few seconds, as if in a dream.

When she woke up, the first thing she thought of was Michael, who would be in the meeting room now. Would he remember what to do? She had to hope he would—she knew full well that reaching him by phone was not an option—and had to think of the rest of the city. They had talked about this event so much and his company had invested more than any other in training its managers, like him. In fact, the major tech companies headquartered in Seattle had invested billions—on top of those provided by the federal government—in preparation for this event: money that went into hiring, training, and equipping rescue teams; into sensors for monitoring the seismic fault and early warning systems that could instantly detect and track the incoming tsunami; and in robots that on a day like this could work side by side with Liz and her team and help them save as many lives as possible. These robots had cut their teeth doing inspection and maintenance work at the tech companies' headquarters and industrial sites, while taking part in all the major disaster exercises, where they would work side by side with dogs and, of course, humans like Liz.

Liz's colleagues were looking at her, waiting for instructions. "Send out the drones immediately," she said. "One-third on the waterfront, sounding the alarm. The other two-thirds all over the city, assessing damages."

The event she had obsessed over for the previous six years had finally happened on her watch. The Cascadia earthquake, the strongest seismic event and the worst natural disaster that can hit the continental United States, had arrived. Overlooked by seismologists up to the 1980s, the Cascadia Subduction zone—the plate boundary stretching from Vancouver to northern California—had been their main concern since the 2010s. By then, experts had come to agree that this fault could cause earthquakes up to 9.0, possibly 9.5 magnitude, much worse than the worst possible shock in California along the San Andreas fault, traditionally the most studied and most feared seismic fault in the country. They also came to agree that such an event was not at all unlikely. In 2010, after examining geological records of past earthquakes in the area, a group of American researchers estimated a 37 percent probability of a magnitude 8 or higher earthquake hitting the Pacific Northwest within fifty years.[2] The last crucial bit that seismologists had agreed on was that the Cascadia earthquake, unlike the San Andreas ones down south, was very much likely to cause a tsunami, similar to the one that devastated Indonesia and other parts of Southeast Asia in 2004. The damage wrought by a megathrust earthquake in a region that had only started to seriously prepare for it in the 2000s was surely going to be huge. But the tsunami would bring the real destruction.

COLLEGE STATION, TEXAS, USA, CURRENT DAY

It is a cliché to say that all people above a certain age remember where they were on September 11, 2001. But for most roboticists, you could probably say the same thing about any major disaster that has hit the Earth over the past few decades: the Twin Towers in 2001, Hurricane Katrina in 1995, the Deepwater Horizon oil spill in 2010, the Tohoku earthquake and tsunami in 2011 and the accident that it caused at the Fukushima nuclear plant; most recently, the COVID-19 pandemic, a very different, but no less tragic kind of, disaster. Each of these events forced scientists and engineers with even a slight interest in robots to think that there must be a better way—a way to find and reach survivors faster; a way to avoid risking the lives of rescuers and firefighters; a way to reach that place at the bottom of the sea where no diver can go; a way to put machines instead of humans in the line of fire. No wonder search and rescue often appears high up on the list of potential applications of most robotic projects. After all, it is the rare example of a job where no one has objections to the idea of robots replacing humans.

In some sense, disasters are a by-product of human technology, so it makes sense to use technology to cope with them. Earthquakes, hurricanes, and volcanic eruptions have hit the Earth since the dawn of time, but their impact depends on what humans have done on the territory they hit. As seismologists often say, earthquakes do not kill people; collapsing buildings do. At some point in history, humans have begun fabricating catastrophes on their own: city fires, train wrecks, explosions at factories and power plants. Indeed, technology has been making disasters more frequent—either directly, by building ever larger transportation systems, infrastructures and factories, or indirectly, with global warming, making floods and hurricanes more frequent.

And yet disaster robotics did not appear as a research field in its own right until 1995. "That was a big year" recalls Robin Murphy, then at the Colorado School of Mines and now one of the world's most renowned experts in disaster robotics. In that year, two very different disasters hit on both sides of the Pacific. In January, the Hanshin earthquake killed more than 6,000 people in Japan. At the University of Kobe, the city that suffered the most losses, Satoshi Tadokoro started a research group that would later become the cornerstone of Japan's International Rescue Systems Institute. In April, a bomb exploded at a federal building in Oklahoma City, killing almost 170 people and injuring over 680. The search for survivors trapped under rubble lasted more than two weeks and motivated Murphy to repurpose her research group and turn her attention to the problem of how to integrate robots into rescue teams in order to speed up the search when the next disaster hit.

Murphy moved to the University of South Florida (USF) in the late 1990s and finally to Texas A&M University a decade later, where she is Raytheon Professor of Computer Science and Engineering. She remains a world's leading figure in disaster robotics. Anyone approaching the field seriously takes more than a quick glance at *Disaster Robotics*, her 2014 book on the subject.[3] Watch her 2018 TED Talk, like more than 1 million viewers already did, and you'll be captivated by her energy and empathy—the qualities you'd hope to find in anyone leading a team through an emergency—as well as by her encyclopedic knowledge of disaster robotics.

For sixteen years, Murphy led the US-based Center for Robot-Assisted Search and Rescue (CRASAR), a nonprofit institute that she helped create and that coordinates the deployment of robots on disaster scenes and acts as an interface for rescue agencies and academic and industrial groups that build and operate robots. September 11, 2001, was pivotal for disaster robotics, she says. A few hours after the planes hit the Twin Towers, CRASAR put together seventeen robots from four different groups, including Murphy's own team at USF. Four small, tracked and wheeled robots were used at the World Trade Center site for penetrating the rubble and reaching the basements and stairwell, hoping to find surviving firefighters trapped in. It was the first documented use of robots at a disaster scene. "Most of the robots used there were already commercially available or under development from the military, but that was the first time the military themselves realized what they could do with them." The use of robots in other disasters had a more mixed success. Hurricane Katrina in 2005 saw the first use of small aerial vehicles. "Unfortunately, it had a negative effect overall," Murphy says. "There were so many groups with so many drones flying around and not thinking enough about the process. In the end, the Coast Guard complained, and the Federal Aviation Authority grounded all drones and made it difficult to fly them for many years." By the time Hurricane Harvey hit in 2017, the lesson had been learned. "Drone use by that point had been routinized, and they were used by the responders in a coordinated manner."

Flying robots are now a standard presence in disaster operations, Murphy notes, but not all modalities (a word used by roboticists to describe how a robot moves) are progressing at the same pace. "There has been a lot of progress in the use of aerial robotics for search and rescue in the last decade, but not so much in ground robots and no progress in marine robotics," Murphy notes. Marine robots are particularly underrated. And yet some of the risks we will face over the next decades are related to water. Extreme events like hurricanes and exceptionally heavy rains are becoming more frequent because of climate change and are more likely to cause floods whose effects are amplified by deforestation and soil erosion. Sea levels are projected to rise in coastal areas, again as an effect of climate change. As for tsunamis, up to December 2004, most people would know them only as the subject of *The Great Wave off Kanagawa*, the most famous work of Japanese painter Hokusai. After the catastrophic wave that hit the Indian Ocean, experts in many countries have started paying more attention to the risk of earthquake-induced tsunamis—especially in the United States, where seismologists have suggested that the West Coast could someday be hit by a tsunami even worse than the one that devastated Indonesia in 2004.

The problem, Murphy says, is that before seeing a family of robots used in a disaster scenario, you need to have a critical mass of them adopted in other routine applications so that when that day comes, emergency operators can trust them. That critical mass is not there for aquatic robots yet. Murphy says that she is "surprised that the transportation sector in the United States has been reticent to adopt underwater robots for inspecting bridges. Bridges have to be inspected every two years, some of them every year. The underwater portion, in particular, requires specialty diving because of strong currents and turbidity. A lot of bridges in the United States are reaching seventy-five years, which should be the end of their life, and they are not getting inspected because there are not enough divers." And yet robots that go underwater can make a big difference. In 2011, Murphy led a CRASAR team that deployed four underwater, remotely operated robots after the Tohoku earthquake and tsunami: "They had four hundred miles of coastline that had been devastated: ports, shipping channels, bridges, and not nearly enough divers." The robots, equipped with cameras and acoustic sensors, inspected bridges and port infrastructures and helped locate and recover bodies. "In four hours we were ready to reopen the Minamisanriku port that would otherwise probably reopen after six months. But even after that, those robots were not put into everyday use at the port."

The years since 9/11, says Murphy, have taught her a couple of lessons that everyone working on rescue robots should keep in mind. First, robots are an extension of human rescuers, not a replacement. "In most cases, the purpose of the robot is to enable responders to sense and act at a distance in an unknown situation," she says. In disaster situations, it's impossible to know in advance what the actual task will be. "We can tell you what the goals are: saving lives. But you can't decide how to do it until you see what is actually going on, and deciding what to do requires expertise. You need someone who knows about bridges, or about drainage, who knows what to look for." The idea of fully autonomous robots reaching a disaster site, finding their way and deciding what they do by themselves, and finally coming out with a few survivors in tow is likely to remain confined to science fiction. Rescue robots could indeed use more autonomy because they often have to operate where remote controllers cannot reach them. But their primary job will remain to go back to human rescuers as fast as possible (or at least to wherever they can get and send a radio signal) to report on what they saw.

The second lesson is that robots need to be tested in routine operations before being used in the frantic hours after a disaster. "Nobody—NOBODY!—believes robots work," Murphy quips. Rescuers will never use the latest, most innovative robot straight out of the lab, no matter how cool it is. "The robots that are used in a disaster are always robots that already exist. People do not have time to learn something new. Rescue managers are accountable. If they use something super-new and strange and it screws up, they lose their job. And in some cases, the robot can actually make things worse. We technologists often think that any robot is better than nothing. That's actually not true. Robots can make things worse. They did at the Pike River coal mine disaster in 2010, where four robots entered the mine but were disabled by an explosion and effectively blocked the only way in and out of the cave."

The third lesson is that human-robot interaction is key. "More than half of the failures of robots in critical operations are attributed to human errors," Murphy says. "But most of the time it was not the operator's error; it was the designer who gave the robot such a bad interface. That is always going to be the biggest barrier to adoption."

SEATTLE, WASHINGTON, USA, SEPTEMBER 26, 2046

11:20 a.m., Nile Headquarters

After trying for a few minutes to open the door with all their might, Michael and Steve had to accept they were stuck on the seventh floor. The earthquake had deformed the door frame to the point that the heavy fireproof metal door to the stairwell would not move no matter how hard they pushed or whatever they threw at it. Elevators were obviously not an option: the whole building was without electricity, and getting into an elevator in the midst of a one-in-a-millennium seismic event would not be a good idea in any case. "Maybe we could just wait here for someone to rescue us," Steve said.

Michael looked out the window and tried to stay lucid while going through it all like it was a long, bad dream. He tried to remember the worst-case scenario that Liz had described to him. The water could flood up to five floors? Six, maybe? He wasn't sure enough to feel safe and just wait there at the seventh floor, but for the moment, there was nothing else to do

Michael and Steve had managed to keep their spirits high so far, but they were now beginning to despair. From their window, they watched the tsunami sweep downtown Seattle and hit their building, the water rising wave after wave, throwing cars, motorbikes, and tons of debris against the windows of the floors below them. They had had a brief moment of relief when they realized that the waves were losing strength and that their floor had been spared by the flood. But then Michael's worst fear materialized: with a terrifying sound, the building bent five degrees on one side: the tsunami wave had made it lose its balance, and the Nile headquarters was now leaning like the Tower of Pisa.

They knew they needed to get out fast, but first they needed to let someone know they were there. Michael reached for the mobile phone in his pocket, hoping that maybe service had been restored, but it was still useless. All they could do was try their best to make themselves heard. They began to shout, then shout and bang on windows and air pipes through the slit between the elevator doors—anything that could transmit some sound and let their voices get out of the deserted, isolated office floor. They had no idea how many people were still in the building, if their colleagues downstairs were safe, and if the ones upstairs had had the same thought and managed to reach the rooftop.

In the end it was something, not someone, that heard them. Michael noticed a few drones hovering around the building and immediately went to the window and started waving his arms, shouting, banging on the cracked glass, hoping the robots' camera and sound sensors would pick up their presence. It took some time, but finally one of the drones perched on the building right at their level and stood on the other side of the glass as if it was indeed looking at and listening to them. Michael and Steve began shouting their names, explaining how they'd remained stuck there. They did not even know if it made sense, if the drone could somehow relay their message to someone—but Michael found comfort in the idea that maybe Liz was somehow listening and watching on the other side.

12:35 p.m., FEMA Headquarters

Liz had seen the tsunami wave hit on her screen through the images provided by the surveillance drones flying over the waterfront. She watched the waters submerge the waterfront, the very same bars and promenade where she and Michael had had breakfast in the sun that morning. Then she saw the wave hit the rest of the city behind, cancelling streets and bridges, lifting cars and trucks as if they were toys, finally hitting the first floors of the skyscrapers downtown. It had been surreal to watch the scene unfold in silence on her screen while trying to imagine the horrific noise it was making a few kilometers away, in the real world. The waves were now losing strength, and the water level was beginning to decline. It was time to act.

On the large screen of the control room, Liz could now see dozens of dots of various colors marking the locations of the robots in different parts of the city. The red dots were drones; there were flocks of them, bird-like robots able to cover large distances and agile enough to fly among buildings and perch on poles and walls. Liz had sent flocks of them downtown and to the most geologically vulnerable areas of the city. Using cameras, laser sensors, and microphones, they were creating tridimensional maps to guide the rescuers. In the meantime, they were also dropping short-range, low-power transmitters and receivers in strategic places, making up for the interruption of mobile phone service and creating special communication networks for the rescuers.

Blue dots were amphibious robots, artificial versions of salamanders and crocodiles that had been immediately deployed to the flooded areas as soon as the waters were starting to recede, to look for people trapped in cars or homes or clinging to floating debris. They were helping Liz and her colleagues check the damage, locate survivors, and look for entry points into damaged buildings but also cutting electric cables and removing debris to make way for the rescue teams. Yellow dots were quadruped robots, used to do the heavy work in buildings and houses: unblocking passages; extinguishing small fires; reaching people; bringing water, medicines, and tools; and communicating with rescuers. They all carried cameras and other sensors allowing them to understand the scene, know if the terrain was too soft, if the water could be traversed. But they also had an arm—actually, something that was halfway between an arm and an elephant's trunk—to bend tubes, open cracks in walls, move aside objects, and cling around pipes to hear human noise throughout the building.

All around Liz in the control room, a dozen of her collaborators were occupied with what to the untrained eye might seem like an avant-garde stage performance. Wearing suits and gloves made out of synthetic fiber, with virtual reality (VR) headsets covering their eyes, they were waving their arms in the air, pointing fingers and dragging around imaginary objects around, climbing invisible steps, spreading their arms and bending on a side like birds do when they change direction midair.

They were in fact remotely controlling some of the robots, the ones that had reached the most challenging spots of the disaster site and could no longer rely on autonomous control alone. The human movements were captured by the sensors on the suits and gloves and turned into instructions, transmitted to the robots' onboard computers. The VR headsets used the images and sounds collected by the robots' cameras and microphones to create a real-time, tridimensional view of the area where each machine was operating. In this way, Liz's colleagues could perceive and move as if they were inside the robots, right on the scene, when in fact they were miles away.

In the end, it was one of these robots that, unbeknown to Liz, was going to the rescue of Michael somewhere on that map.

Search and rescue is often advocated by academic researchers as a promising application of mobile robots. Postdisaster emergencies present rescuers with unexpected situations that require multiple physical and cognitive skills. However, in contrast to the multitasking abilities of human beings, most robots are designed for a single task: drones for flight, legged robots for moving on uneven terrain, and manipulators for picking up and placing objects. However, humanoid robots can walk, manipulate, drive, and even fly airplanes, although today it is still rare to see humanoids excelling at more than one task or even performing two tasks at the same time. Could those anthropomorphic robots bring the multitasking capabilities required in search-and-rescue missions? To answer that question, in June 2015 the Defense Advanced Research Projects Agency (DARPA) brought together several humanoid robots in Pomona, California, to test their abilities in disaster-mitigation situations. The DARPA Robotics Challenge was endowed with a $2 million award for the humanoid that could autonomously complete in the shortest time several tasks that are typical of a postdisaster scenario: drive to a building and get out of a vehicle, reach for a door, turn a door handle and enter the building, locate and turn off a valve, climb stairs, disconnect and reconnect a cable to a different socket, grab a screwdriver and drill a hole through a wall. The winning robot, Hubo from the Korean Advanced Institute of Science and Technology, took approximately forty-five minutes to complete the tasks, which nevertheless had to be greatly simplified during the preparatory competitions. While for roboticists this was a great achievement and coronation of several months of hard work, a human rescuer would have probably taken less than ten minutes to accomplish the same tasks. The runner-up robot, an Atlas humanoid built by the company Boston Dynamics and programmed by the Florida Institute for Human and Machine Cognition, took only five minutes longer than Hubo but eventually tipped over while waving at the enthusiastic crowd.

The challenge was depicted in the popular media as a failure, and more than 2.5 million people watched on YouTube the compilation of tragicomic falls of the robots. But this was no failure. The DARPA Robotics Challenge and the efforts of the best computer scientists and engineers around the world vividly showed that accomplishing those tasks, even in a simplified form, requires physical and cognitive capabilities that are much more diverse and sophisticated than driving a car in traffic, something that artificial intelligence has not yet completely mastered. But things are progressing fast. Since June 2015, Boston Dynamics, the company that developed the Atlas humanoid robots used by many teams in the DARPA competition, kept releasing newer versions capable of walking in snow, getting back on their feet after a fall, lifting heavy boxes, and even performing somersaults. While it is difficult to assess the capabilities of those machines in real-world scenarios because the company does not reveal details of how those robots operate, academic research in humanoid robotics has been making progress on all fronts, and we are confident that by the end of this century, humanoid robots will be working alongside humans and replacing them in dangerous tasks, including search and rescue.

Meanwhile, a team of roboticists from Swiss universities and start-ups, including our lab at Ecole polytechnique fédérale de Lausanne (EPFL), is taking a different approach to search and rescue. Instead of betting it all on a single complex robot, Swiss roboticists are working on a multirobot rescue system composed of several specialized robots: drones, legged robots, amphibious robots, and distributed artificial intelligence that enables them to share information with each other and with humans. They argue that this "Swiss army knife" approach, where the most appropriate robotic tool is deployed according to situational needs, could be more robust to failure, better promote and leverage progress in simpler robots, and give professional rescuers more flexibility in gradually adopting the technology. The research program, sponsored by the Swiss National Science Foundation, does not offer a multimillion cash award but gives researchers the opportunity to work with rescue teams and participate in training events where humans and robots work side by side in simulated rescue operations that feature collapsed buildings, fires, flooding, and stranded persons. For researchers, this experience is humbling and precious at the same time—humbling because we realize that our prototypes are still quite far from being used in the real world and precious because we learn where robots are most needed in postdisaster emergencies and how human rescuers prefer to use them.

What does it take for aerial and ground robots to join search-and-rescue missions? Let's start with drones. Today, most commercial drones come in two shapes: they have fixed wings—just like those of commercial airliners, only much smaller—or helicopter-like propellers, with the most common configuration being a quadcopter lifted by four propellers. Just like their bigger and peopled cousins, airplanes and helicopters, each drone type has its pros and cons. For an equivalent weight, fixed-wing drones can fly faster and longer and are easier to pilot, build, and maintain than multicopters. These characteristics make them ideal for reaching people who are stranded in areas that cannot be rapidly reached by ground vehicles or for collecting precious information at remote locations that are dangerous for humans to approach, such as nuclear facilities or erupting volcanoes. Multicopters, however, are more versatile than fixed-wing drones: they can fly vertically, accelerate and slow down to a stop in the air, turn on a dime, and need less space to take off and land. These characteristics make them ideal for landing in confined spaces, such as through a small opening in a forest or on the top of a building, and for continuously monitoring an object of interest without moving. Some drones have flapping wings, similar to those of insects or birds. Flapping-wing drones display an agility that almost matches that of multicopters, but they are still energetically inefficient and fragile compared to their fixed-wing and rotary-wing cousins, and therefore they are still found mainly in research laboratories. It will take major leaps in the power efficiency of electrical motors, or the invention of novel artificial muscles that can compete with the endurance of propeller-driven drones, for flapping-wing drones to take off from labs and offer a competitive advantage with respect to fixed-wing and rotary-wing drones, but when it happens, they will combine the best of both worlds.

In our laboratory at EPFL, we try to bring together the advantages of different drone types into a single morphing robot that can change shape and locomotion mode. In nature, many animals can modify their body shape to move more efficiently in different environmental conditions.[4] For example, vampire bats fly and land next to their prey, where they use the tips of their wings to silently move on the ground and jump on their prey. The ability to fly over large distances and move on the ground could be useful for reaching the area of a chemical spill or nuclear fallout and performing accurate inspection on the ground. Our bat-inspired drones have retractable wings whose outer edges can rotate.[5] In the air, the drone flies like a conventional winged drone and uses the wing tips to change direction; on the ground, it retracts the wings to take up less space, shifts the center of mass to improve balance, and uses the rotating wing tips as spiked wheels to move around on irregular terrain. Another version consists of a quadcopter with spiked wheels at the two extremities of a central tube holding four foldable arms.[6] Once landed, the drone folds in half and relies on the spiked wheels to rapidly move on the ground while dragging the propeller arms on its back, thus offering rescuers unprecedented views of small spaces inside a collapsed building. Once the inspection mission is completed, the drone unfolds its rotor arms and flies back to the operation center.

Wing morphing is also useful during flight. For example, birds of prey spread out their large wings to produce large aerodynamic forces required for rapid turns and decelerations, thus performing agile maneuvers, but they keep the wings closer to the body when they want to fly fast in a linear trajectory in the open sky, reducing aerodynamic resistance and energy consumption. Similarly, our avian-inspired drones have wings and tails made of artificial feathers that can fold like those of birds, to fly over larger distances, withstand strong winds, perform sharper turns around buildings, and fly at slower speed without stalling.[7] Another morphing drone developed with colleagues at the University of Zurich consists of a quadcopter with foldable arms and vision-based stabilization.[8] The drone can shrink in flight to pass through a narrow hole into a building, retract the front propellers to get closer to objects for inspection, and even use the arms to grab objects and release them at another location.

When it comes to ground robots, all those that have ever been used so far in disaster operations move on track and wheels, but they have difficulties moving on rough terrain. Think of the many ways a car can get stuck.

It cannot climb stairs, it gets easily trapped in mud, and it has a hard time overcoming obstacles or gaps—problems that most animals that walk on four legs do not have. Alas, moving on legs happens to be a hard problem in robotics, one that cannot be approached with good, old-fashioned control strategies, where every movement of the robot is precisely programmed in a predetermined sequence. Each hole in the ground, each additional slope degree, each moving stone can make the robot lose its balance and change what the next "right" step would be.

Marco Hutter's lab at ETH Zurich has created a quadruped robot that addresses locomotion in creative ways, taking advantage of the spectacular success that machine learning techniques have had in the past few years. Named ANYmal—a reference to both its bioinspiration and its ambition to walk anywhere—it is an agile robot that can climb slopes and stairs, step over obstacles, cross gaps such as ditches and ducts, and duck under hanging obstacles. It has cameras and laser sensors and all the computational power it needs to autonomously explore and map unknown environments.

To make ANYmal learn to walk, Hutter resorts to a clever mix of simulation and real-world experiments. Roboticists typically build a simulated version of their robots on a computer to test control methods before deploying them on the robots. Unfortunately, developing accurate simulation models from first principles turns out to be extremely challenging. For example, joint actuators do not perfectly follow commanded signals due to bandwidth limitation, friction or saturation effects, and communication delay. Hutter's group proposed a new method for combining simulation models with data collected from the sensors of the real robot to close the gap between simulation and reality. His simulator generates the massive amounts of data required to train a neural network to produce gaits that are effective when deployed in the real robot, including on very challenging terrain.[9]

Together with Luca Gambardella in Lugano and other teams around the world, Hutter's group endowed his quadruped robots with machine learning capable of predicting terrain properties based on earlier experience.

Currently ANYmal is being upgraded with arms that will allow it to manipulate things while it walks—for example, open a door, activate a switch, or turn a dial.

As Murphy notes, it takes a critical mass of adoption in other applications, where robots can be used on a day-to-day basis, before rescue operators can trust them enough. Meanwhile, the ETH spin-off ANYbotics has started to commercialized the ANYmal quadruped for industrial inspection. Their latest product is deployed in diverse situations ranging from offshore wind farms, oil and gas sites, over large-scale chemical plants and construction sites, to sewers and underground mines to autonomously conduct visual, thermal, acoustic, and geometric inspection tasks.

One day, a robot such as ANYmal could be a game changer in disaster situations, in particular when buildings, factories, and power plants are hit. But as our fictional story set in Seattle shows, and as Robin Murphy notes, many disaster situations require moving not on soil but in water or, worse, to transition between water and land in the half-submerged environments that are common in the aftermath of a flood, hurricane, or tsunami. There, the dog-inspired ANYmal could get stuck.

What would really be handy are robots that can effortlessly move in water and on the ground like amphibians, which is what our colleague Auke Ijspeert has been studying and building for years at EPFL in Lausanne. Amphibians and reptiles use what zoologists call "sprawling" gait: their four legs extend laterally instead of vertically right under the body, like mammal legs. Wave-like lateral movements propagate through their elongated bodies while they walk, a reminder of their aquatic origins (it is how eels and other snake-like fish swim) and a secret weapon that lets these animals transition effortlessly between land and water. Their very low center of gravity makes them also stable on muddy, irregular, and uncertain terrain.

At his lab in Lausanne, Ijspeert has spent many years building robots that recreate their gait—robots that could one day be used to inspect submerged infrastructures like bridge pillars or intervene in flooded areas or buildings. "The goal," he says, "would be to send them where a rescue dog cannot go, either because it is too dangerous—dogs, like human rescuers, should not risk their lives more than it is necessary—or because a dog cannot physically walk there."

Ijspeert's first robotic salamander was a modular robot, made up of a series of interconnected units, corresponding to the vertebrae of the actual animal.[10] The robot's brain triggered a pattern of lateral movements that flew from one joint to the next, making the spine oscillate horizontally at different frequencies for swimming and walking. When on the ground, four short limbs lifted the robot just enough above the ground to walk, still moving its spine left and right in the snake-like fashion that is typical of amphibious creatures. *Salamandra robotica* was followed by a more complex robot called Pleurobot, with more joints and sensors and the ability to climb steps in water.[11] In 2015, Ijspeert was approached by the BBC. The broadcaster's filmmakers wanted a robot that could mix unnoticed with crocodiles in Africa and help filming them. Ijspeert's team studied the morphology of crocodiles and giant lizards, using photographs and many visits to Lausanne's Natural History Museum. Thus, the K-Rock robot was born. Its articulated skeleton had three joints in the head and neck, two in the spine, four in each leg, and three in the tail. Videos and x-ray recordings of the actual animals allowed Ijspeert and his team to fine-tune the motion controller. Covered in a waterproof, crocodile costume, K-Rock made for a great camera operator for the BBC series Spy in the Wild, allowing the team to shoot amazingly close takes of adult and baby Nile crocodiles.

K-Rock's future job, however, is expected to be that of a search-and-rescue operator in flooded areas. Compared to its amphibious predecessors, it is lighter and more powerful, meaning that it can lift itself higher above the ground and also carry additional payload. In addition, it is portable, not an unimportant feature for a disaster-scene robot. It weighs less than 5 kilograms, and its parts can be disassembled and fitted into a trolley-size package. After successful tests in the lab's pool in 2020, K-Rock is now able to transition between walking, crawling, swimming, ducking under a 15-centimeter-high passage and squeezing through a narrow corridor. But there is still a lot of muddy ground to cover before rescue managers could send it in a collapsed and flooded building to look for survivors.

"The sensing part is still largely missing" Ijspeert admits. "For robots to walk and swim like animals, you need to measure the load force on the body and on the skin, the interaction forces, and they are very hard to measure. And then the big challenge is the control part. How do you close the loop between sensors and motors? How do you perform locomotion on different grounds? How does the robot decide when to switch from swimming to crawling and walking, and how does it choose the best gait for a given terrain?"

Robots will gain increasing intelligence, autonomy, and locomotion capabilities in the years to come, but humans will continue to play a supervisory role in search-and-rescue missions and intervene when necessary. So far, the quality and quantity of data coming from the robot's video cameras and other sensors have been greatly limited by temporal delays and the low bandwidth of wireless networks and Internet protocols, making real-time teleoperation quite difficult. In the near future, 5G wireless communication networks will make it possible to receive video and sound data with a maximum delay of 1 millisecond and enable control of robots in real time. But as Robin Murphy noted, the interface between robots and humans will be crucial to establish an intuitive and informative bidirectional communication. Today, most telerobotic interfaces consist of keyboards, video displays, touch screens, or remote controllers, which require training to reduce the risk of human error in mission-critical situations.[12] For example, conventional drone controllers are not intuitive and are cognitively demanding in long-term operation.[13] Natural body gestures could be a more intuitive method to control robots than joysticks and keyboards. To this end, the Swiss robotics team has developed a range of wearable technologies that detect natural body motion and translate it into robot commands. Researchers at the University of Lugano focus on situations where the human and the robot are within line of sight.[14] Their smart wristband with embedded artificial intelligence lets humans summon drones and legged robots with the same gestures used to instruct dogs. The human points at a drone and makes it land by indicating a desired location on the ground, or it instructs a quadruped robot to search in a desired direction by pointing at it and shifting the arm to the search location.

In our laboratory at EPFL, we instead focus on situations where robots are not in sight. Our goal is to develop wearable technologies and software that let humans perceive and behave as if they were the robot, somewhat like in James Cameron's *Avatar* movie, where humans enter a symbiosis machine that connects their brains and bodies to those of flying animals. But we want to go one step beyond and make the symbiosis machine fully wearable. In collaborations with colleagues in neuroscience, we performed a series of experiments in virtual reality to understand how humans would fly with their body (if they could).[15] Based on those data, we developed a soft exoskeleton, which we call the FlyJacket, that conveys the video feed, wind sound, and aerodynamic forces from a drone to the human and translates human body motion into flight commands for the drone.[16] Yves Rossy (also known as Jetman), the first man to fly with jet-powered wings strapped on his back, read news of the FlyJacket in the media and asked to try. As he smoothly moved his torso, head, and arms from the comfort of a stool in our lab, Yves told us that he felt like being in the air and used the same body motion when flying in the sky with his jet-powered wings. Since then, hundreds of people have tried the FlyJacket at public events, and all could fly without any prior training, even if they had never piloted drones or played video games. Most people spread out their arms and use the torso to steer the drone, while turning their head to explore the surroundings through the frontal camera placed on the drone. Contrary to conventional remote controllers, the FlyJacket does not require the use of hands. Therefore, we developed smart gloves that let the wearer touch thumbs and fingers to place pins on a map at locations seen from the drone. A rescuer could use the FlyJacket to survey a disaster area with a drone and place red pins to indicate locations with people in need of help, blue pins to indicate flooded regions, and green pins to indicate safe zones—all of this done precisely and in real time by a single person without the typical cognitive effort required to control a drone.

We are currently working on machine learning methods to make the symbiotic control experience even more personal and tailored to different body gestures, motion styles, and morphologies.[17] We believe that wearable technologies and machine learning, combined with 5G connectivity, will create a novel form of symbiosis between humans and intelligent machines and dramatically augment human operation at remote locations through robotic bodies.

The Swiss team is not alone in developing new robots for search and rescue and for disaster operations. A few years ago, the European project ICARUS developed a concept of a heterogeneous robotic team combining air, ground, and marine vehicles with ad hoc communication networks.[18] It included two unpiloted ground vehicles with tracked locomotion; an autonomous marine capsule capable of navigating on the water surface; drones of various sizes; and mapping tools that combine input from the aerial and ground robots. Another European project, TRADR, focused on human/robot teams in disaster response scenarios, in particular tracked robots. This human/robot team was deployed and used for inspecting damaged buildings after the 2016 earthquake in central Italy.[19]

On the other side of the Atlantic, at Harvard University, Professor Robert J. Wood has been working on insect-sized flying robots for over a decade. His tiny robots, which are called RoboBees but are in fact more closely inspired by flies, first proved capable of controlled flight in 2013 and have since then acquired other skills such as swimming, perching, and pivoting mid-air.[20] Weighing only around 100 milligrams, they have thin, insect-like wings attached to a vertical structure: imagine a matchstick with wings on the top, and four tiny legs for support on the bottom. One of their most impressive feats is the frequency at which their wings can flap—more than 120 times per second. To achieve such fast movement, Wood's team could not use the electromagnetic motors that power most robots; although they are extremely effective, they are not easy to scale down to such a miniscule size. Instead, the team developed piezoelectric actuators based on ceramic strips that change shape when an electric current is run through them.

Another challenging part of the design was finding ways to control the RoboBees's flight. For drones that light and small, even a breath of wind can be a destabilizing force. Wood's team developed algorithms run by an off-board computer that could control each wing independently, carefully balancing the speed of upstroke and downstroke wing movements to keep the RoboBee stable and flying in the desired direction. A modified version of RoboBees with some additional components, including a tiny buoyancy chamber, could dive into water, switch from flying to swimming, move toward the surface, jump out of the water, and resume flying.[21] Another version could perch on a surface thanks to electroadhesive materials applied to its four tiny feet.[22] And yet another could hover and pivot in mid-air as well as land precisely on a spot the size of a one-cent coin.[23]

A swarm of such tiny and highly controllable robots, each capable of moving around different environments and carrying sensors to perceive their surroundings, could find many applications in disaster management in addition to aiding pollination in agriculture, thus fulfilling Wood's original vision.

In the private sector, the US company Boston Dynamics has developed several quadrupedal robots for military and civilian applications, including BigDog, Spot, and Spot Mini, along with the humanoid Atlas. Their YouTube videos regularly make a splash and show impressive performance, from Atlas back-flipping to Spot Mini opening a door by grabbing and pulling a handle with an extendable arm. However, there are no publications that describe their systems and evaluate their performance, an understandable choice for a private company.

Between 2019 and 2020, another DARPA challenge—this time focused on subterranean environments—pushed the envelope of robotic autonomy in extreme environments even further.

In the urban scenario, an unfinished nuclear plant in Washington State, robots had to climb stairs, move through narrow corridors with sharp bends, and inspect holes; they gained points by correctly identifying carbon dioxide emitters and hot devices scattered throughout the course. The robots had to work autonomously for most of the track because radio signals from the engineers could not reach them down there. The winner in that case was

COSTAR, a team led by the NASA Jet Propulsion Laboratory, whose goal is to design a team of autonomous robots to search for signs of life in subsurface voids on other planets. For DARPA's urban circuit, COSTAR adapted two Spot Mini loaned by Boston Dynamics and fitted them with a bespoke control unit and algorithms to make their way through the dark hallways and up and down the stairs.

The Swiss team, led by Marco Hutter, ended up winning the final event of the DARPA subterranean grand challenge, in which robots were tasked with traversing underground tunnels, an infrastructure similar to a subway, and natural caves and paths with rough terrain and very low visibility, all the while searching for objects and reporting their locations to a control station. The team combined four ANYMals and caged drones derived from our lab at EPFL, scoring 23 "points" (i.e., identifying 23 of the 40 objects placed along the course). "These competitions are a nice way of enforcing benchmarking in a field where it is very different to replicate what other research groups are doing, because everyone is using different machines," Hutter comments. "Also, what I like about these challenges is that there is a lot of uncertainty until the very last minute. You have to go through very different locations with no prior knowledge, which forces you to remain generic. And that is one of our key problems in robotics: How can we generalize so that our robots can solve different problems, even the ones we can't anticipate?"

International challenges have been crucial for the development of autonomous cars, and it is a safe bet that they will have a role in shaping the future of rescue robotics too: pushing platforms, algorithms, and human-machine interfaces to their limits, getting them to the point where they could save lives—many lives—in the aftermath of a disaster such as the one we have imagined in this chapter. Even more important is signaling society's interest in this field of robotics and forcing everyone—researchers and policymakers alike—to consider it a priority. Unlike in other stories we tell in this book, the building blocks of the technologies we are imagining are already there for the most part. What is missing, and what can make the difference in the future, is the political will to invest in technologies for disaster preparedness.

SEATTLE, WASHINGTON, USA, SEPTEMBER 26, 2046

3:40 p.m., Nile Headquarters

Michael and Steve were safe now, but they knew they would not forget those two hours for the rest of their lives. Stuck on the seventh floor, with the building dangerously leaning on a side, they had almost lost hope of getting out alive when they heard the sound of bent metal coming from the elevator area. They saw the two doors part, then a mechanical arm appeared between them and fully opened them. Finally, an agile four-legged robot hopped inside the office and ran toward them. As they would later learn, rescuers had been alerted by the perching drone that someone was trapped on their floor. After an amphibious robot had been sent to explore the ground floor and find the safest—and driest—way up, this sturdiest and faster quadruped had been sent in and had made its way up the inclined stairs, removing debris with its trunk, eventually reaching them through the elevator cavity after realizing the door from the stairs was blocked.

Once on the office floor, the robot presented them with a package it carried on its back. Inside they found water, energy bars, and a complete first-aid kit—welcome gifts, since the floor's water tank had fallen on the floor and broken during the earthquake, and Steve had cut his hand while trying to open the door to the stairways. Being in a leaning building with the ever-present risk of seismic aftershocks was still far from reassuring. But now they knew there was a way out.

Finally, a group of firefighters arrived on the building top by helicopters guided by the images and 3D map constructed from data sent by drones, amphibian, and legged robots. They fast-roped down to Michael and Steve's floor, used a hammer to crack the window open, and finally reached the two men. Taking them all up by ropes would have been too dangerous and, in the end, would have taken longer. So the firefighters had the robot pulling the deformed metal door with its trunk, until it bent enough to let them pass, and escorted Michael and Steve to the stairs. While climbing, they were joined by other colleagues rescued from the floors above, until the bandwagon reached the building top. There, one by one, they were lifted up to helicopters that would bring them to the sports arena that FEMA

had converted into a shelter for the homeless and survivors—following the emergency plan that Liz had reviewed, revised and rehearsed so many times before that day.

SEATTLE, WASHINGTON, USA, SEPTEMBER 2, 2047

Being here again on this summer morning, with almost the same warm breeze they so vividly remembered from a similar day one year earlier, sent shivers down Liz and Michael's spines. But they had to be there. The reopening of the waterfront bar where they'd spent the morning before the earthquake meant so much for the city, which was finally beginning to recover, and for the two of them as well.

Half of the waterfront was still wasteland, its buildings and piers inaccessible and badly in need of repair. In the weeks after the quake, Seattle's residents, volunteers from all over the United States and Canada, and armies of robots from laboratories and rescue agencies from every continent had joined forces to get the worst-hit areas of the city back on their feet. The work was far from over. Swarms of drones were still flying back and forth over downtown Seattle, restoring the external walls that had been stained by mud and repairing structural damage. Dozens of wheeled robots were still at work in various parts of the city removing the remaining debris. It would still take at least one or two years before the waterfront could get back to what it once was. The bar was not expecting too many clients for the moment, but reopening it was a sign that the city was beginning to put the tragedy behind.

So were Liz and Michael. For a long time, it had been almost impossible for them to think of anything else, no matter how hard they tried. Both were overwhelmed by work in the months after the earthquake. Liz had to coordinate operations until every survivor was located, every victim was identified, and every building's safety was assessed—all the while giving talks and witnessing at congressional hearings on the handling of the emergency. Michael also had briefly become prime time news material, being interviewed over and over about how he was rescued, before being put in charge of a new research program at his company that would improve and expand the use of robots for disaster mitigation.

Two months before that September day, they had discovered that they were expecting a child. Their first reaction had been utter fear: Were they prepared for a baby? For a few days, they had considered moving to another city and starting anew, in a place less fraught with memories and away from the risk of new disasters. But they decided to stay. "We can't run away from risk," Liz had told Michael one morning. "If there's one thing that day has taught us, it's that you just have to prepare for it."

3 OUR FIRST MARTIAN HOMES

As she approached the surface of Mars and prepared to execute the landing maneuver she had rehearsed nearly a hundred times in simulation, Leila knew what to expect. For years during her training, she had been studying every detail of every photograph that past space missions have taken of Northeast Syrtis—the area in the northern hemisphere of the planet chosen as the ideal settlement for the first human colony on the Red Planet. Then, during her nine-month flight, ground control had been sending her the pictures from the camera-equipped rovers that were roaming over that area of the planet. Week after week, those images showed her and the rest of her crew how that small patch of Martian land was being transformed in preparation for their arrival. Still, when she finally got there, the view left her "breathless," as she later wrote in a message sent to the control center right after landing.

Millions of people who were watching the live stream of the landing were also holding their breath—albeit with a thirteen-minute delay, the time it took for the transmission to travel from Mars to Earth. Against the reddish background of the Martian soil stood a dozen tall, tree-like structures, their wide leaves gently swayed by the winds on the planet's surface. From a distance, it could look like an oasis, one of those green areas that grow around a water source and interrupt the dullness of our Earth's deserts—except, of course, nothing was alive on Northeast Syrtis—at least, not before the arrival of the Mayflower II mission that Leila was leading. Those branches and leaves belonged to plant-inspired robots, brought to Mars about eighteen months earlier by an unmanned spacecraft. Since then, those robots had explored, dug, grown in size, captured energy from sunlight, and prepared the ground for the astronauts' arrival. It was thanks to them that, unlike Neil Armstrong and Buzz Aldrin when they first arrived on the moon in 1969, the first women and men to arrive on Mars would have somewhere cozy to stay.

Bringing a human crew to Mars after decades of preparation had been no small feat. But the most difficult part was just beginning: the crew was going to spend six months there, something no human had ever done before, while completing a long series of experiments and exploration missions. At the end, they would perform the first ever takeoff from Mars and withstand another months-long journey back home.

When preparations for this mission began, more than ten years before that landing, a couple of things quickly became clear. The first was that astronauts would need at least some basic life support: huts with walls and roofs that would shelter them from wind and radiation, beds, toilets, heating at night and cooling. The second was that the construction of those huts could not be fully planned in advance. The composition of the soil and the irregularities of the surface were not entirely predictable. Moving sand dunes could change the landscape significantly between planning and realization. Programming rovers to execute a predetermined series of steps was not a viable option, and the thirteen-minute communication delay would make remote control clumsy and inefficient.

Luckily, as early as 2008, a group of Italian researchers had developed a feasibility plan for the European Space Agency (ESA) with a novel idea: deploy plant-inspired robots, which they called plantoids, that would grow roots into the Martian soil, extract energy and water, and create the infrastructure necessary to build habitable shelters. Almost forty years and a lot of research later, plantoids were welcoming the first human colony on Mars.

Unlike most robots we're familiar with, including the various Martian rovers that have visited the planet since the 1970s, moving around was not the plantoids' strongest suit. They had been parachuted on Mars some months earlier by a probe. After a controlled descent aided by rockets attached to the parachutes, the plantoids anchored themselves to the soil and got to work. They unfolded large photochemical leaves to recharge their batteries and began growing artificial roots deep into the soil, centimeter by centimeter, slowly winning the resistance of the rocky Martian soil.

Over the final few months before the landing, the plantoids focused on two jobs. First, their roots sampled the Martian soil, mainly to extract water, which represents around 2 percent of the surface, though bound to other chemicals and not easily accessible. This water then filled reservoir tanks that the astronauts could use. Second, they anchored themselves more and more firmly to the soil, until they became stable enough to support structures capable of resisting the relentless winds blowing on the surface of the Red Planet. They thus became the cornerstones and foundations of the huts that Leila and her colleagues would call home for the next year or so.

PONTEDERA, ITALY, CURRENT DAY

By the early 2010s, the idea of bioinspired robots was no longer a novelty. Engineers had already created robotic versions of salamanders, dragonflies, octopuses, geckos, and clams—a diverse enough ecosystem to allow the *Economist* to portray it in summer 2012 as "Zoobotics."[1] And yet Italian biologist-turned-engineer Barbara Mazzolai managed to raise eyebrows when she first proposed looking beyond animals and building a robot inspired by a totally different biological kingdom: plants. As fluid as the definition of the word *robot* can be, most people would agree that a robot is a machine that moves, but movement is not what plants are famous for, and so a robotic plant might at first sound, well, boring.

But plants, it turns out, are not static and boring at all; you just have to look for action in the right place and at the right timescale. When looking at the lush vegetation of a tropical forest or marveling at the colors of an English garden, it is easy to forget that you are actually looking at only half of the plants in front of you. Admittedly, they may be the best-looking parts but not necessarily the smartest ones. What we normally see are the reproductive and digestive systems of a plant: the flowers and fruits that spread pollen and seeds and the leaves that extract energy from sunlight. But the nervous system that explores the environment and makes decisions is in fact underground, in the roots. Roots may be ugly and condemned to live in darkness, but they firmly anchor the plant and constantly collect information from the soil to decide in which direction to grow for finding nutrients, avoiding salty areas, and preventing interference with roots of other plants. They may not be the fastest diggers, but they are the most efficient ones, and they can pierce the ground using only a fraction of the energy that worms, moles, or manufactured drills require. Plant roots are, in other words, a fantastic system for underground exploration, and Mazzolai proposed creating a robotic version.

Mazzolai's intellectual path is a case study in interdisciplinarity. Born and raised in Tuscany, in the Pisa area that is one of Italy's robotic hot spots, she was fascinated early on by the study of all things living, graduating in biology from the University of Pisa and focusing on marine biology. She then became interested in monitoring the health of ecosystems, an interest that led her to get a PhD in microengineering and eventually to be offered by Paolo Dario, a biorobotics pioneer at Pisa's Scuola Superiore Sant'Anna, the possibility of opening a new research line on robotic technologies for environmental sensing.

It was there, in Paolo Dario's group, that the first seeds—pun intended—of her plant-inspired robots were planted. Mazzolai got in touch with a group at the ESA in charge of exploring innovative technologies that looked interesting but still far away from applications, she recalls. While brainstorming with them, she realized space engineers were struggling with a problem that plants brilliantly solved several hundred million years ago.

"In real plants, roots have two functions," notes Mazzolai. "They explore the soil in search of water and nutrients, but even more important, they anchor the plant, which would otherwise collapse and die." Anchoring happens to be an unsolved problem when designing systems that have to sample and study distant planets or asteroids. In most cases, from the moon to Mars and distant comets and asteroids, the force of gravity is very weak. Unlike on Earth, the weight of the spacecraft or rover is not always enough to keep it firmly on the ground, and the only available option is to endow the spacecraft with harpoons, extruding nails, and drills. But these systems become unreliable over time if the soil creeps, provided they work in the first place. For example, they did not work for Philae, the robotic lander that arrived at the 67P/Churyumov–Gerasimenko comet in 2014 after a ten-year trip. The robot failed to anchor at the end of its descent, bouncing away from the ground and collecting only a portion of the planned measurements.

In a brief feasibility study carried out between 2007 and 2008 for ESA, Mazzolai and her team let their imagination run free and described an anchoring system for spacecrafts inspired by plant roots. The research group also included Stefano Mancuso, a Florence-based botanist who would later gain fame for his idea that plants display "intelligent" behavior, although of a completely different sort from that of animals. Mazzolai and her team described an ideal system that would reproduce, and transfer to other planets, the ability of Earth plants to dig through the soil and anchor to it.

In the ESA study, Mazzolai imagined a spacecraft descending on a planet with a *really* hard landing: the impact would dig a small hole in the planetary surface, inserting a "seed" just deep enough in the soil, not too different from what happens to real seeds. From there, a robotic root would start to grow by pumping water into a series of modular small chambers that would expand and apply pressure on the soil. Even in the best-case scenario, such a system could only dig through loose and fine dust or soil. The root would have to be able to sense the underground environment and turn away from hard bedrock. Mazzolai suggested Mars as the most suitable place in the solar system to experiment with such a system—better than the moon or asteroids because of the Red Planet's low gravity and atmospheric pressure at surface level (respectively, one-third and one-tenth of those found on Earth). Together with a mostly sandy soil, these conditions would make digging easier because the forces that keep soil particles together and compact them are weaker than on Earth.

At the time, ESA did not push forward with the idea of a plant-like planetary explorer. "It was too futuristic," Mazzolai admits. "It required technology that was not yet there, and in fact still isn't." But she thought that others beyond the space sector would find the idea intriguing. After transitioning to the Italian Institute of Technology, in 2012, Mazzolai convinced the European Commission to fund a three-year study that would result in a plant-inspired robot, code-named Plantoid.[2] "It was uncharted territory," says Mazzolai. "It meant creating a robot without a predefined shape that could grow and move through soil—a robot made of independent units that would self-organize and make decisions collectively. It forced us to rethink everything, from materials to sensing and control of robots."

The project had two big challenges: on the hardware side, how to create a growing robot, and on the software side, how to enable roots to collect and share information and use it to make collective decisions. Mazzolai and her team tackled hardware first and designed the robot's roots as flexible, articulated, cylindrical structures with an actuation mechanism that can move their tip in different directions. Instead of the elongation mechanism devised for that initial ESA study, Mazzolai ended up designing an actual growth mechanism, essentially a miniature 3D printer that can continuously add material behind the root's tip, thus pushing it into the soil.

It works like this. A plastic wire is wrapped around a reel stored in the robot's central stem and is pulled toward the tip by an electric motor. Inside the tip, another motor forces the wire into a hole heated by a resistor, then pushes it out, heated and sticky, behind the tip, "the only part of the root that always remains itself," Mazzolai explains. The tip, mounted on a ball bearing, rotates and tilts independent of the rest of the structure, and the filament is forced by metallic plates to coil around it, like the winding of a guitar string. At any given time, the new plastic layer pushes the older layer away from the tip and sticks to it. As it cools down, the plastic becomes solid and creates a rigid tubular structure that stays in place even when further depositions push it above the metallic plates. Imagine winding a rope around a stick and the rope becomes rigid a few seconds after you've wound it. You could then push the stick a bit further, wind more rope around it, and build a longer and longer tube with the same short stick as a temporary support. The tip is the only moving part of the robot; the rest of the root only extends downward, gently but relentlessly pushing the tip against the soil.

The upper trunk and branches of the plantoid robot are populated by soft, folding leaves that gently move toward light and humidity. Plantoid leaves cannot yet transform light into energy, but Michael Graetzel, a chemistry professor at EPFL in Lausanne, Switzerland, and the world's most cited scientist,[3] has developed transparent and foldable films filled with synthetic chlorophyll capable of converting and storing electricity from light that one day could be formed into artificial leaves powering plantoid robots. "The fact that the root only applies pressure to the soil from the tip is what makes it fundamentally different from traditional drills, which are very destructive. Roots, on the contrary, look for existing soil fractures to grow into, and only if they find none, they apply just enough pressure to create a fracture themselves," Mazzolai explains.

The plantoid project has attracted a lot of attention in the robotics community because of the intriguing challenges that it combines—growth, shape shifting, collective intelligence—and because of possible new applications. Environmental monitoring is the most obvious one: the robotic roots could measure changing concentrations of chemicals in the soil, especially toxic ones, or they could prospect for water in arid soils, as well as for oil and gas—even though, by the time this technology is mature, we'd better have lost our dependence on them as energy sources on planet Earth. They could also inspire new medical devices, for example, safer endoscopes that move in the body without damaging tissue. But space applications remain on Mazzolai's radar.

Meanwhile, Mazzolai has started another plant-inspired project, called Growbot. This time the focus is on what happens over the ground, and the inspiration comes from climbing trees. "The invasiveness of climbing plants shows how successful they are from an evolutionary point of view" she notes. "Instead of building a solid trunk, they use the extra energy for growing and moving faster than other plants. They are very efficient at using clues from the environment to find a place to anchor. They use light, chemical signals, tactile perception. They can sense if their anchoring in the soil is strong enough to support the part of the plant that is above the ground." Here the idea is to build another growing robot, similar to the plantoid roots, that can overcome void spaces and attach to existing structures. "Whereas plantoids must face friction, grow-bots work against gravity," she notes. This new project may one day result in robot explorators that can work in dark environments with a lot of empty space, such as caves or wells.

But for all her robots, Mazzolai is still keeping an eye on the visionary idea that started it all: planting and letting them grow on other planets. "It was too early when we first proposed it; we barely knew how to study the problem. Now I hope to start working with space agencies again." Plant-inspired robots, she notes, could not only sample the soil but also release chemicals to make it more fertile—whether on Earth or on a terraformed Mars. And in addition to anchoring stuff, she envisions a future where roboplants could be used to grow full infrastructure from scratch. "As they grow, the roots of plantoids and the branches of a growbot would build a hollow structure that can be filled with cables or liquids," she explains. This ability to autonomously grow the infrastructure for a functioning site would make a difference when colonizing hostile environments such as Mars, where a forest of plant-inspired robots could analyze the soil and search for water and other chemicals, creating a stable, anchored structure complete with water pipes, electrical wiring, and communication cables: the kind of structure astronauts would like to find after a year-long trip to Mars.

MARS, JULY 2055

After switching off the spacecraft engine, Leila took a long, deep breath and let it all out. She screamed at the top of her lungs, so loud that she could barely hear her crewmates doing the same, all of them screaming their joy to be alive. They had made it to Mars. It would take thirteen minutes for the ground control room, their families and friends, everyone on Earth to know that they were okay and almost half an hour before they could hear, in return, the cheering and applause from the control room. Leila decided to relax some more seconds before initiating the long and painstaking procedure that would allow her and the crew to get out of the spacecraft and set foot on Mars. There would be checks to do on the spacecraft, decontamination to avoid bringing bacteria to the planet, special suits to wear. But for the moment, Leila just looked out at the shelters where they would all soon be getting some rest.

They were not five-star hotels, but they looked fine: three constructions, each one about 20 square meters and little more than 2 meters high, irregularly shaped but approximately round, and vaguely resembling the adobe huts that were once common across Africa, Central America, and the Middle East. For the whole duration of their trip to Mars, Leila had followed the construction of the huts—looking at the pictures transmitted by the Mars rovers and suggesting modifications as one would do when supervising the renovation of a home.

A few months into the trip of the Mayflower, after the plantoids had finished their job and prepared the foundations of future human habitats, another squad of robots set out to work and started building walls and roofs for the huts. The construction team included a few big-wheeled robots equipped with bucking drums that they used to dig up regolith, the crushed volcanic rock that lies beneath Mars's easily recognizable, iron-rich red surface layer.[4] Tons of regolith were extracted, stripped of the water molecules and oxygen bubbles trapped in it, and pressed into bricks that were then piled up in the same area where the plant-inspired robots were growing.

When enough bricks were ready, a band of smaller building robots entered the scene and started piling them up to build the huts. Though they did follow a blueprint, a lot of design choices had to be made on the spot. No matter how many rovers have been on Mars and how many times probes have orbited on it shooting pictures, engineers could never know enough about the specific conditions on the planet surface to program every single step of the robot construction and dictate with millimeter-level precision where every brick would have to go. Furthermore, the Martian surface could unexpectedly change at any time because of soil being deposited by dust storms or eroded by winds. The Martian robotic builders had to make up their plans on the go, and for that they relied on control algorithms inspired by termites—some of nature's most ingenious engineers, that decades earlier had attracted the attention of both entomologists and engineers.

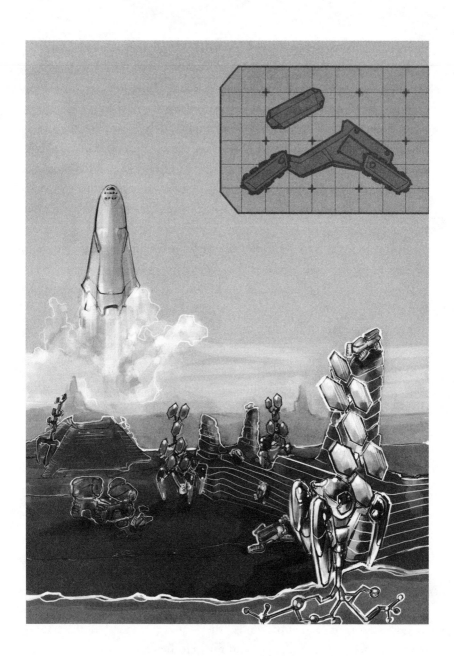

Radhika Nagpal's path to swarm robotics is somehow complementary to Barbara Mazzolai's: whereas Mazzolai is a zoologist who has chosen robots as a tool of her trade, Nagpal is an electrical engineer and computer scientist who turns to living things as a source of inspiration for solving dauntingly complex computational problems. She builds systems made out of several simple and dumb units that can self-assemble into shapes or collaborate to build structures, without any central control unit and using only a few rules of local interaction. The idea, once again, is to make robots more powerful, robust, and *scalable*—a word roboticists love, which means that the complexity of a problem does not grow exponentially with the number of robots involved. In other words, once a solution is found, it holds true for ten as well as for a thousand robots.

A lot of Nagpal's work is based on the observation of how cells, ants, and termites solve their own construction problems. Bioinspiration did not come naturally to her. "As an undergrad, I avoided all biology classes," she recalls, laughing. "I was never too interested in it. But I was very interested in distributed systems: how to create algorithms so that a problem that you would not know how to solve on a single computer can be solved by many computers, and much faster."

By the time she began graduate school, Nagpal became involved with a small group at MIT—which included Tom Knight, Harold Abelson, and Gerald Sussman among others—that started a project on "amorphous computing." "Our idea was that in order to build the best parallel computer, we had to look at the best parallel computer in nature, and that is parallel behavior," she explains. Do not just think of animals: you can find parallel behavior in ants but also in cells or even in molecules and particles that fluctuate. "Statistical mechanics is a beautiful example of parallel behavior that appears in unreliable, stochastic pieces," she continues. "It was a beautiful vision, and cells were the first examples I gravitated toward, and in my PhD thesis, I explored the idea of a foldable material made of cells that would deform, with a few origami-like rules allowing the material to take different shapes." At the time, in the early 2000s, it was a visionary idea that would be realized only several years later by scientists such as Daniela Rus at MIT and Jamie Paik at EPFL, to mention only two.

Later, upon setting up her own lab at Harvard University, Nagpal began working on robotic self-assembly—small units coming together to form a larger structure, like cells do, or robots collaborating to build something much bigger than their bodies. For her first attempts at creating robotic builders, she turned to a low-cost solution: "I am a computer scientist. I didn't really know how to build things at first. So with my students, we began making robots out of Lego blocks. It was the 3D printing of the mid-2000s, a way to quickly assemble what you needed for experiments. And it taught me a lot. I realized how hard building robots is, but also how many new ideas you get from building them."

When it comes to construction, there is no better inspiration in nature than termites. It is hard not to stand in awe at these tiny insects' capability to build amazing structures. They are not the only builders among insects, but they are the most ambitious. Their nests—or mounds—rise up to 10 meters or more, adorned with chimneys and pinnacles. Some species, such as the compass termite, are able to align their mounds along the north-south axis so that the internal temperature increases rapidly during the morning while avoiding overheating in the afternoon, a textbook example of thermal efficiency. Entomologists such as J. Scott Turner, a world-renowned authority on termites, have wondered for a long time how the nest master plan is encoded in the collective mind of the colony.[5]

After years of studying the structure of the mounds, filming their construction, and performing computer simulations of termite behavior, researchers have come up with an elegant and simple explanation.[6] Each insect follows only a few simple, genetically regulated, rules of local interaction, such as, "If the termite before you does A, you do B; if it does C, you do D." And they implement them by relying on something researchers call *stigmergy*:[7] they use their environment as a communication medium, each termite leaving behind on its path small clues that the next insects use to make a choice. In finding the shortest route between food and nest, termites follow tracks with higher concentrations of chemical traces left by other termites. As these chemicals evaporate over time, the shortest path, which is traversed more frequently within the same amount of time, will maintain the highest chemical concentration and attract more insects. In nest construction instead, the behavior (deposit material, move on, dig further, remove material) is dictated by the state of the structure that the termite encountered. For example, if the material deposited at a location exceeds a given height, the next termite will deposit its cargo at a different location; if a tunnel is shorter than a given length, the next termite will keep digging; and so forth. All of these behavioral rules, which resemble if-then computer instructions, are genetically encoded in each termite. There is no master architect, no project manager, no supervisor. Even the largest and most complex nests result from thousands of insects following such simple rules, a textbook case of collective intelligence emerging from local interactions.

Nagpal decided to try transferring the same capacity to robots but with an additional twist: not only a swarm of robots would build a complex structure following simple local rules, but humans would be able to choose à la carte what shape the resulting "nest" would end up with. Such a robotic swarm could, for example, quickly build a shelter, a barrier, or even a dam in a disaster situation. "The other basic concept that we borrowed from termites, besides stigmergy, is that they build something, then they climb over it and build some more." Whatever they build, in other words, immediately becomes a platform to build something more, as well as a cue on what to build next. "We could not make the robots climb over straight walls, as termites do, so we made them build staircases and climb over them."

The project, called Termes, resulted in a paper that made the cover of *Science* in February 2014, in which Nagpal and her coauthors, Justin Werfel and Kirstin Petersen, proved that a team of small robots, guided only by stigmergic rules, could autonomously build a complex, pyramid-like structure made of bricks that robots picked from a stack.[8] Though not as small as termites, her robots were quite simple affairs, equipped with *whegs* (a contraction of wheels and legs to describe an array of leg-like rods emanating from a rotating central hub), infrared sensors to detect black-and-white patterns, an accelerometer to know whether they are climbing or descending, an arm to lift and lower bricks, and ultrasound sonar units to measure distance from the "building" and from other robots.

Nagpal and her team provided the robots with a picture of the desired structure, and the algorithm would convert it into movement guidelines for robots at each location—in a sense, simple traffic laws suitable for that structure: how to move around, climb, and from which side to descend. From then on, each robot would just pick a brick, circle around the structure until reaching a special brick that would act like a point of reference, climb onto the structure along any path allowed by traffic laws, and attach its brick at the first vacant spot where the geometry of the surroundings satisfies some predefined conditions. Then it would go back, pick a new brick, and start it all over again. Each individual robot "knew" only the rules that specify where a new brick can be attached and where it cannot. By laying a brick, the robot changed the appearance of the structure and thus influenced the next robot's behavior (there goes stigmergy). None of the robots knew what the others were doing, at what stage the construction was, or even how many other robots were there. In fact, the system worked pretty well regardless of the number of robots; it was, in other words, very scalable.

In 2014, shortly after presenting their termite-like robots, Nagpal's group made headlines again with another study, bringing self-assembly in robot swarms to a whole new level. This time, the robots were the bricks, and there were a thousand of them.[9] More precisely, 1,024 small "kilobots" could self-assemble into different configurations, such as into a star shape or into the letter K, without predefined plans and without oversight. The researchers would simply provide a graphic representation of the desired shape; four robots would then position themselves at key spots, thus marking the origins of a coordinate system, and all the others would use local interactions and simple rules, such as "stay within a given distance relative to the coordinate system and to other robots." Seen from above, their collective movement resembled how birds flock across the sky. The algorithm Nagpal designed ensured that hundreds of these robots could complete tasks based on high-level instructions provided by humans, a step that Nagpal considers essential for the future of distributed robotics. "Increasingly, we're going to see large numbers of robots working together, whether it's hundreds of robots cooperating to achieve environmental clean-up or a quick disaster response, or millions of self-driving cars on our highways," she says. "Understanding how to design 'good' systems at that scale will be critical." Chosen by *Science* as one of the top 10 scientific breakthroughs of 2014, the kilobots have since been developed into an open-source robotic platform that other researchers are using to experiment on self-assembly and self-configuration.[10]

Until the publication of the Termes article, Nagpal and her team had to teach themselves about termites the same way we would all do: by reading entomology papers. It was only after their *Science* cover article that they started collaborating with biologist J. Scott Turner and his group. "In this kind of research, you really have to convince the biologists to collaborate with you," she explains. "If all you do is math, they have no reason to take you on a field study. But once we had that project going and we could show that our robotic model was addressing some of the same questions Turner was studying on the field, we were in a much better position to go to him and basically say, 'Please, let us come with you! Let us see for real how the natural system works.'"

Nagpal and her team joined Turner for several research trips to Namibia, where termites build some of the most impressive and most photographed architectural masterpieces. "We wanted to see termites at work to study how emergent properties actually emerge," she explains. "How do they sense air quality to decide what they are going to build? What are the cues they pick up from the environment? How do they know if they are building a roof or a column?" Nagpal's team contributed their expertise with video tracking experiments, where the movements of individual insects are captured on digital cameras and analyzed with mathematical tools in search of rules that govern their behavior. Turner made Nagpal reflect on the weaknesses of her initial approach. "One particular weakness he noticed was that we had such a rigid system, not easy to reconfigure when something does not go down as expected," she recalls. "And that had a lot to do with our choice of materials." Real termites have a big advantage over her robots: they use a soft material, mud, that allows them to correct errors. "If I misplace a Lego block, I can't put the next one in place. But if I'm building with mud, I can always shove some more mud in. The material allows termites to be sloppy and imperfect." Sloppy and imperfect: that is what construction robots tasked with building things autonomously and in unpredictable environments would need to be.

"Those trips to Namibia changed the course of the next projects we would work with," she recounts. With Nils Napp, then a postdoc in her lab and now at SUNY in Buffalo, she started devising robots that can build with amorphous materials. Their first attempt was a wheeled robot, in the form of a little car, that could drive on rough terrain. On the front, they mounted a printing head that sprayed polyurethane foam, a material that can quickly expand and become solid. The fact that it expands so much means the robot is capable of building quite large structures while carrying around only a small bucket of foam. At Harvard, Nagpal and Napp also experimented with robots that can throw toothpicks and glue them with foam and amass sandbags to create vertical structures that can work as a levee. After becoming a professor at SUNY Buffalo, Napp went on to a new step: robots that can use whatever they find on the scene to build structures. They search for rocks, take them on the building site, and put them together using a small amount of gluing material they carry with them. In 2019, Napp managed to demonstrate how a group of robots with different construction capabilities used different materials to collectively build a structure, choosing themselves from time to time what was the best material to use for the next "brick in the wall."[11]

As they progress, Nagpal's little builders and self-assembling robots could have applications in disaster response or in large self-assembling structures, such as the dams we've imagined for the fictional Venice in chapter 1. But just like for the plant-inspired robots, it is in planetary exploration—where the environment is the most hostile and unpredictable and instructions from Earth cannot be used to control robots in real time—that this kind of growing and self-assembling robots could one day prove their worth.

MARS SURFACE, 2055

"Okay, guys, keep your passports at hand—it's time to check into the hotel."

It had taken the crew almost an hour to go through the rigid script of checks and procedures to complete before disembarking and after ticking the last box on her tablet, Leila could finally smile and go off script. The whole crew went along with her smile. As they prepared to set their feet on Martian soil, walk a few hundred meters, and enter the huts where they would make themselves at home, they resembled a group of tourists readying to get off the tour bus and drag their luggage to the hotel reception desk—except, of course, they were all wearing space suits and were about to visit a place no other tourist had visited before.

The crew lined up behind Leila as she pressed the button. They waited for the hatch to open, then proceeded to take the walk they—along with countless other inhabitants on Earth—had been dreaming about for so long. As they marched toward the hut, they absorbed the sight of the Martian landscape—simultaneously so familiar, thanks to the hundreds of images taken by rovers, yet so shockingly new to human eyes.

It took them about five minutes to reach the huts, where a small band of termite robots had gathered to form a welcome committee, a little surprise that Houston ground control had arranged for them.

They entered the main hut and looked around. The structure looked reassuringly solid and more spacious than they had expected. The temperature was surprisingly comfortable: a web of pipes, stemming from the plant-inspired robots that supported the structure like pillars, ran along the interior walls and on the floor. The sunlight captured by the robotic leaves was used to heat a few liters of water that constantly circulated in the pipes, keeping the place at an average 20°C.

For all their efficiency, the robots had not been able to do everything by themselves. Soon Leila and her team would have to install power lines and outlets to harness the electricity generated by the robots and distribute it at various points in the huts, allowing the crew to charge all the tools they would need for their experiments. There was also some more work left to add partitions and booths inside the hut to provide a bit of privacy needed for the six months. But overall, the astronauts could only thank the robots for providing them with homes on a planet no human had ever before called home.

4 DRONES AND THE CITY

HONG KONG, 2036

The buzz of her phone forced Jane to take her eyes off the computer screen for the first time in two hours. With the deadline for submitting her project only a few hours away, she had hardly allowed herself to even look out of the window the whole morning, but the alert flashing on her screen told her that she now had to. She got up from her desk and gazed at the view from her fifty-sixth-floor office in Wan Chai. Out there, the city sprawled across the Victoria Harbour into Kowloon and was as busy as ever. Everywhere was rush hour—down there in the streets and out there on the water across the two town districts, as well as up there, hundreds of feet above the ground and beyond the top of the skyscrapers that surrounded her.

Traffic noise from the ground could hardly reach her at that height—not least because noisy combustion engines were becoming a rarity in technology-minded megacities like this one. And yet the sky outside her window was all but silent. Drones were all around, their buzz making the whole upper part of the cityscape sound like a giant beehive. Flying vehicles of all sorts—some with helicopter-like rotary wings, others flying or gliding on light wings—were crisscrossing the sky, headed to deliver everything from packages to baskets of delicious dim sum to urgent medical supplies.

A professional architect, Jane could not help but marvel at how much drones had changed the way cities were designed—and, hence, her job. In the early twentieth century, elevators had allowed cities to start growing in a third dimension but had left the job undone. People could live and work in tall buildings, but all movement of goods between them still happened on the ground. Plus, building and maintaining those skyscrapers still required humans to perform dangerous tasks that resulted in accidents, sometimes fatal ones. By the 2030s, drones had finished the job, taking a lot of traffic away from the lower parts of the city and at the same time allowing for better, safer, and more regular maintenance on the tallest buildings. What Jane was witnessing here in Hong Kong was happening in New York City and Lagos, in Tokyo and in São Paulo, in Shenzhen and Moscow.

She watched thousands of drones as they zipped back and forth between buildings, over and below bridges, flying side by side and crossing each other's path without ever colliding or bumping into walls, and always finding their way to destination. It looked easy now, but it had taken decades of work to make drones safe, reliable, and sufficiently intelligent to change the megacities of the twenty-first century.

ZURICH, SWITZERLAND, CURRENT DAY

Over the past decade, drones have taken the world by storm. The history of unpiloted aerial vehicles (the technical term engineers use for them) is actually much longer, with the first prototypes dating as far back as World War I. In 1918, one year after the United States had entered the conflict, a secret collaboration began between Orville Wright and Charles Kettering. Orville was the elder of the two brothers who invented the first airplane and the only one surviving at the time, after Wilbur's premature death in 1912. Charles was a prolific inventor who would spend more than a quarter-century as head of research for General Motors. Together they designed the "Kettering bug," still on display at the National Museum of the Air Force in Dayton, Ohio. This twelve-foot-long wooden biplane, powered by an internal combustion engine, was launched by a sort of slingshot moving along the rail and was not meant to come back. Its mission was to carry a bomb and crash onto its target, the kind of mission that some years later, in World War II, would fall on Japanese kamikaze pilots. The control system of that early drone—which never got to be used in battle, although fifty aircrafts were built and ready to roll when the conflict ended in November 1918—was quite simple. Before launch, operators would use the target distance and direction and wind speed to calculate how many cycles of the engine were needed for the drone to reach the target. After the last scheduled cycle, presumably while flying over the target, a cam would be released by a clockwork-like mechanism, detaching the wings and effectively turning the aircraft into a missile.

It would take a few more decades before remote control via radio frequencies could be used to guide unpiloted aircrafts in their flight and guide them back to base after the mission. Drones then played their part, and an increasingly important one, in military operations, their technology becoming increasingly more sophisticated from World War II to the Vietnam War, from the Yom Kippur conflict up to the Afghanistan and Iraq wars in the twenty-first century. But it was in the second decade of this century that drones crossed over to the civilian world, finding peacetime applications and becoming consumer products.

Civilian drones are much smaller than military drones: they cover shorter distances, carry smaller payloads, and must be safe when flying in proximity to humans. The miniaturization and commercial availability of electronic components, such as on-board computers, batteries, and sensors, largely driven by the personal device industry, has accelerated research and development of small aerial machines. Smart drones can now be bought at prices comparable to that of a smartphone.[1] Today they are routinely used for documenting how forests, coastlines, and rivers change over time and filming endangered animal species without disturbing them—not too much, at least. They are popular in agriculture, where they carry sensors to monitor crop growth, spot pests and parasites, and provide instructions to automated tractors that fertilize the fields. Drones are also used for security, from monitoring borders to critical infrastructures such as power plants, and they are increasingly used in search-and-rescue operations, as we saw in chapter 2. Drone footage has become a common tool in filmmaking. And drone delivery is currently being experimented with parcels where speed of delivery is vital: medical samples and blood bags. However, drone operations today require a human operator, or are restricted to especially authorized regions if the drone has some degree of autonomy. In particular, drones are not yet allowed to fly autonomously between buildings and in close proximity to humans because most of them do not have the perception and intelligence of a human pilot. These drones may collide with undetected obstacles and break in pieces on the ground or, worse, cause injuries to humans. However, several possible solutions to these challenges are being investigated in research labs around the world.

Davide Scaramuzza, at the University of Zurich recently achieved impressive results in making drones fly autonomously and at high speeds using vision alone, dodging moving obstacles and even performing acrobatic maneuvers. Cameras, he says, are not the perfect sensor for robots but offer the best trade-off of size, weight, and performance.

"If you ask me what the ideal sensor for a mobile robot is, I have to answer that it is a laser," says Scaramuzza, who as a PhD student cut his teeth in the field of driverless cars, where laser sensors called LIDARs do the lion's share in making vehicles perceive and "read" their surroundings. "Lasers can see everything, also in the dark. But they are too heavy and require too much power to be used on a small drone."

Whereas the first experiments with autonomous ground vehicles date back to the 1980s, it took more than twenty additional years for vision-based drones. In order to move autonomously between two locations, a drone needs at the very least to know its own position in space and the position of its destination. In principle, GPS could do that. But it can easily become unreliable when something stands in the way of the satellite signal, for example, between tall buildings, and just stop functioning indoors, underground and in buildings with thick walls. Cameras instead always work—provided there is at least some light.

Scaramuzza's first vision-based drones used a technique called visual SLAM (for simultaneous localization and mapping), where the drone's onboard computer does four things with the images coming from the camera. First, it extracts key features to identify structures and build a map of the surroundings, searching for lines, shapes, and light gradients that are maintained from one camera frame to the next, and are likely to be walls, roads, poles, or buildings. Second, it calculates the drone's position on that map, using the apparent movement of those same key features as an indication of how distant they are at any time from the drone. Third, it uses all that information to plan a path in the environment, a suitable route for getting the drone from A (where it is now) to B (where it wants to be—for example, a building worth exploring). Fourth, it executes the path, giving the rotors the right sequence of commands to go through all the points in space that make up the planned route.

"This is the so-called rule-based approach," Scaramuzza explains. "Each block has to be written by a coder who sets out rules that the drone's computer must follow to make decisions—rules about the environment, about how the actuators work, about physics." As accurate as these rules can be, they cannot account for all the uncertainties in a complex, real environment other than a laboratory setting or a limited outdoor test track.

Today it is hard to find a sector in robotics or computation that is not being transformed by artificial neural networks, but autonomous, vision-based flight is certainly among the most affected. Scaramuzza has been using a particular flavor of machine learning, called "imitation learning." "The idea is to teach a neural network to fly a drone by imitating an 'expert.' This way you do not need to know all the rules that regulate the environment and the machine." The only "rule" is "fly as similar as possible as the 'expert,'" leveraging the expert's knowledge without having to learn it from scratch in a formal way. "The 'expert' is not a human demonstrator but rather an algorithm that has access to privileged information, such as the true attitude and velocity of the drone and perfect knowledge of the environment and surrounding obstacles. Using such information, the 'expert' can calculate the optimal behavior of the drone."

Rather than flying a real drone for many hours—which, especially when performing aggressive or even acrobatic maneuvers, would risk damaging and even losing some robots, Scaramuzza uses simulations, like roboticists often do in the early stages of their projects. The drone and the surrounding environment are simulated on a computer, and the expert flies this simulated drone in order to collect examples that are then fed to a neural network. The network then uses those examples as a starting point to conduct many simulated flights, and in the process it learns to fly by itself.

"The biggest problem with robotics," Scaramuzza notes, "is that rule-based approaches generally work very well in controlled environments, but it's impossible to anticipate how all real environments could look like. However, neural networks are very good at generalizing their acquired knowledge."

The "expert" drone flight becomes a mathematical function of the visual information gathered from the onboard camera, and the neural network learns over several attempts to approximate that function. In other words, it learns to fly the simulated drone as the expert would do. The last step is what roboticists call sim-to-real transfer: when the performance in simulation is solid enough, the neural network is promoted and is used to control a physical drone flying in the real world.

Overall, it works surprisingly well. "By replacing rule-based control with learning, we managed to fly drones with a robustness we never had before. We can fly them in forests, or in drone racing courses, reaching speeds up to 60 kilometers per hour," says Scaramuzza.

Scaramuzza sees speed as a key requirement for future drones—not because they need to be spectacular but because they need to be more efficient. "Batteries have not improved enough, and currently are the main bottleneck for drone operations," he notes. In the best case, a small quadrotor can fly for thirty minutes before its battery runs out. At their current low speed—the only speed they can afford today without losing stability and bumping into obstacles—that time may not be enough to cover a useful distance and deliver a parcel or inspect a building during a search-and-rescue mission. "Maybe one day fuel cells will provide more energetic autonomy for a comparable mass, but right now, the only alternative is to increase speed and accomplish more in less time," he says.

In order to make his drones faster, Scaramuzza relies on a new family of cameras that depart from the typical logic of electronics and more closely mimic what happens in the nervous system of humans and animals. "Neuromorphic cameras have been on the market more or less since 2010, and we were the first ones to use them for drones back in 2014," Scaramuzza says. "They are great for detecting fast-moving things, because they work as a filter that responds only to the information that changes over time." Their key advantage, for drones in particular, is their low latency, which is the time it takes for the camera to give feedback when something is moving in front of it. For a standard camera, latency can range from 1 to 100 milliseconds. For an event camera, it is between 1 and 100 microseconds.

Traditional video cameras work by activating all their pixels to take snapshots of the whole scene at a fixed frequency, or frame rate. This way, though, a moving object can be detected only after the onboard computer has analyzed all of the pixels. Neuromorphic cameras have smart pixels that work independently of each other, similar to what individual photoreceptors in the insect and human eyes do. The pixels that detect no change remain silent, while the ones that see a change in light intensity (also called an "event") immediately send out the information to the computer. This means that only a tiny fraction of the all pixels of the image will need to be processed by the onboard computer, therefore speeding up the computation a lot.

Since event-based cameras, as they are also known, are a recent innovation, existing object-detection algorithms do not work well with them. Scaramuzza and his team had to invent their own algorithms that collect all the events recorded by the camera over a very short time and subtract the events due to the drone's own movement, which typically account for most of the change in what a normal camera would see. The few pixels that are left correspond to moving objects in the field of view.

By equipping a quadrotor with event cameras and tailored algorithms, Scaramuzza and his team were able to reduce the drone's reaction time when presented with an obstacle, which would normally be between 20 and 40 milliseconds, to just 3.5 milliseconds. In a study published in 2020 in *Science Robotics*, Scaramuzza and his team flew a drone indoor and outdoor while throwing objects at it. The drone was able to avoid the objects—including a ball thrown from a 3 meter distance and traveling at 10 meters per second— more than 90 percent of the time.[2]

It turns out that deep learning and event cameras make a perfect marriage. "For the first three or four years we used the new cameras with classic algorithms in order to just understand how the device worked," says Scaramuzza, "but we are now applying learning to them as well, which allows us to do something extremely difficult: modeling the noise from the event camera." That is something that engineers need to do with every sensor: they need a mathematical model of how the electronic noise can affect the sensor's signal. But event cameras, which operate like biological neurons, are highly nonlinear, meaning that their behavior is not easily predicted by simple mathematical rules. "The way we used learning to overcome the problem was to train a neural network to reconstruct a full image in grayscale from the signal of the event camera." By themselves, event cameras are great for detecting movement and sending digital instructions to a computer, but not really for reproducing an image: their output signal is not something you can easily save as a jpeg file and view on a digital screen. But Scaramuzza and his team trained a neural network to do just that. It's only in black and white, but it works. "We did it in simulation first, and then transitioned to a real camera. It now works so well that you can watch a real video through an event camera. And because it can adapt to a very large range of lighting conditions, you can point the camera toward the sun and see everything, whereas a normal camera would be blinded by the light." This camera also has a very high temporal resolution, which Scaramuzza demonstrated by filming garden gnomes being shot with a rifle. Being able to reconstruct a slow-motion video of the gnomes' explosion convinced the scientific community that they were really onto something. "This is something you normally do with large, expensive, high-speed cameras. We did it with a small camera, much cheaper, that uses less data and can store hours and hours of footage." Today, major chip manufacturers produce and sell neuromorphic chips for applications in vision processing and parallel computation. This trend will most likely reduce chips' dimensions and price, and make them reliable and affordable enough for deployment in small commercial drones.

The next challenge for Scaramuzza will be to keep driving down the latency in his visual sensors, making drones increasingly more reactive and making their algorithms better at interpreting a scene where everything moves fast. His laboratory's goal is to develop algorithms that do away with the old, compartmentalized treatment of the four control blocks—mapping, localization, path planning. and execution—and treat the four problems the way a living brain would treat them. "Working out these things separately and in sequence is not what humans and other animals do" he notes. "We want our algorithms to do all four things in parallel and close the loop between perception and action."

Drone visual autonomy, he says, will also require developing better simulation environments and better techniques to transition to the real world. The challenge lies in being able to generalize whatever is learned in simulation to many different environments, not only to the one on which the simulation was actually based. "We want to figure out how to generalize across different domains," he explains. "For example, training a drone on a simulated race course, but making sure it learns things that can be useful also when inspecting a building."

How will we know when scientists have cracked autonomous flight? A good indication would be when a drone flying on its own beats the best human pilots on the race course. "Drone racing is proving a great way to demonstrate technologies, attract research groups, and solve complex problems," says Scaramuzza, who regularly participates with his group in large drone racing challenges. His team has shown that a neural network paired to the drone's control unit can be trained on a simulated racetrack and then successfully apply what it has learned to a real racetrack, even if the gates are partially hidden or camouflaged or when they are moved in different positions from one lap to the next.[3]

In drone races, autonomous drones now manage to fly at substantial speeds through complex tracks but still finish five or six seconds behind the best human pilots. But Scaramuzza is confident that drone racing will have its "AlphaGo" moment at some point, when an autonomous machine will leave even the best human pilot in the world behind—just like Alphabet's neural network in 2016 crushed Lee Sedol, the Go game world champion, convincing even skeptics of the rising power of neural networks and deep learning.

HONG KONG, 2036

The phone informed her that it was time to go: the drone would land in a few minutes. Jane left her room, headed to the elevator, and pushed the button for the top floor, where she got out on the rooftop and into the humid morning air. Looking south, where her package was coming from, she noticed a small band of quadcopters cleaning windows at various floors of the high-rise just in front of her own; some meters below, another one was repainting a section of the outside wall. In the distance, Jane could see other drones doing routine tests on the whale-shaped roof of the exhibition and convention center. Eventually, she noticed a drone heading straight toward her. As it approached, it slowed down, retracted the arms with the propellers into a protective grid, and paused for a few seconds right in front of her, cross-checking her face and ID badge. Then it landed next to her on the inductive recharging pad and finally spoke: "I have a parcel for you, Jane."

She picked up the box and began to carefully open it while the drone recharged its batteries on the plate. In a few seconds, she held in her hands the 3D printed parts of the 1:500-scale model of a pedestrian bridge, the missing piece she needed to complete the project for showcasing at city hall the following day. "Parcel accepted," she said with a smile and instructed the drone to leave as soon as it was ready. "Thank you Jane," the drone's voice replied. "I will be ready for departure in three minutes and twelve seconds. Have a nice day."

Jane got back to her office where the 3D printed bridge reached its final destination. With this latest addition, the model that had been taking shape for weeks on her big table finally showed every detail of how she and Michael, the head of another architectural firm on the Kowloon side of the city with whom she hoped to win the bid, were planning to redesign Nathan Road, one of the city's busiest streets, clogged by slowly moving cars and people lining up at bus stops and walking by the numerous shop windows.

Most citizens, even those most familiar with the area, would have had trouble recognizing any sign of the current configuration of the street in this plastic model. Indeed, Jane's and Michael's project aimed to turn what once was one of the most trafficked areas of the city into a vast pedestrian area with bike lanes, fountains, and gardens. It was a dramatic change, but the city planners' instructions were clear: now that traffic had moved upward, the ground floor had to go back to the people! Throughout the 2030s, drones had taken so much traffic off the roads that entire parts of the city could now be redesigned, taking car traffic out of the equation. Drones were now the default mode for exchanging parcel and mail, in large cities as well as in the suburbs. For that to happen, during the 2020s and early 2030s, drones had

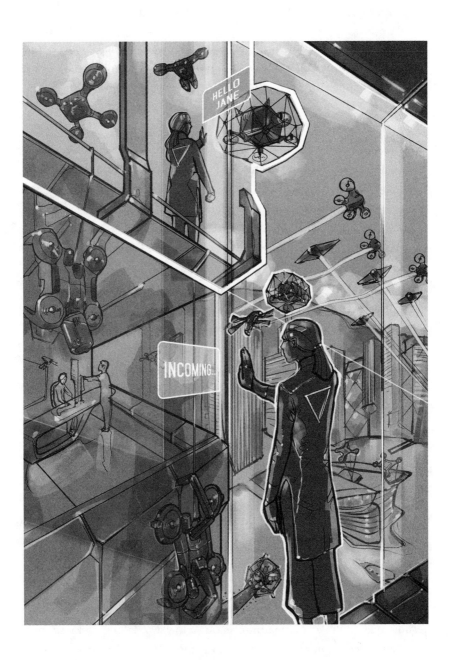

to become not only faster and more autonomous than their predecessors, but also safer for people to handle and more resistant to collisions—which sometimes happen, no matter how good the control system is.

LAUSANNE, SWITZERLAND, CURRENT DAY

Even sophisticated vision systems, such as the ones Scaramuzza and other researchers around the world hope to develop, may fail to see electricity cables, may be confused by light reflections, or may fly into transparent glass. It turns out that flying insects face similar challenges and do indeed collide with their surroundings. However, instead of falling to the ground and breaking up as today's drones do, which would amount to death, insect bodies can absorb strong impacts thanks to flexible exoskeletons, flexible compound eyes, and foldable wings that deform and rapidly spring back in place after an impact. These observations led us to rethink how drones are designed and give them collision resilience. Our goal is to make robots capable of operating in the physical world by recasting difficult problems of perception and artificial intelligence into smarter mechanical design and materials. Instead of relying on complex visual and control algorithms to avoid impact, our collision-resilient drones can withstand physical contact and resume flight. At EPFL in Lausanne, we are investigating two collision-resilience strategies, both inspired by insects.

The first strategy consists of encapsulating drones within flexible exo-skeletons, functionally similar to those of insects.[4] Our drones are protected by a mesh of lightweight carbon-fiber rods interconnected by flexible rubber-like joints. These exoskeletons let the air flow through the propellers and can absorb the impact of collisions by slightly deforming around the joints. In-air collisions can destabilize the drone and make it fall to the ground. In early prototypes, we added retractable legs for uprighting after a fall and resume flight, similar to what insects do.[5] However, falling and resuming flight after a collision requires significantly more energy than staying in the air. Therefore, we went back to the lab and redesigned the protective exo-skeleton as a rotating cage, similar to gimbal desk globes, that freely rotates around the core propulsion system. When the drone hits an obstacle, the exoskeleton absorbs the impact and rotates without affecting the attitude of the propulsion system, thus allowing the drone to stay in the air and resume its trajectory. These exoskeletal drones could slide against walls, roll on the ground, and autonomously traverse a dense forest by relying only on a magnetic compass.[6] A video camera positioned on the inner propulsion system gave us impressive footage of the scene as the drone came into contact with trees, rolled over bushes, and steadily progressed toward the northern edge of the forest. In 2015, this collision-resilient drone won the United Arab Emirates Drones for Good competition for being the only robot capable of locating a person in a collapsed building located across a water stream. Former students in our lab used the $1 million award to start the company Flyability and commercialize a richly sensorized version of this drone for the inspection of confined spaces, such as large industrial boilers, oil tankers, elevated bridges, and underground tunnels.

Exoskeletons have disadvantages, though; they add weight and aerodynamic drag and thus reduce flight duration compared to a naked drone of the same size. Our second strategy for making drones collision-resilient consists of rethinking the structure and material of their bodies instead of building protective structures around them. During flight, the drone arms on which the propellers are mounted must stay rigid in order to resist aerodynamic lift forces; however, in the event of an impact, the structure must deform to absorb the significantly larger forces and quickly regain the original configuration to resume flight. It turns out that insects have solved this problem by evolving foldable wings with elastic joints. These wings can resist the aerodynamic forces generated during flapping without folding but give way under the stronger forces generated by a physical impact, and rapidly snap back in flight configuration after the collision.[7] Inspired by these studies, we started to design drone structures made of flexible materials, such as thin polymer foils folded along strategically located lines.[8] These collision-resilient drones can be destabilized by strong collisions and fall to the ground, but they do not break. The choice of one of the two strategies (a collision-absorbing cage or a resilient naked structure) depends on the application. If the drone is expected to experience several collisions or fly in close proximity to a structure, an exoskeleton is the more suitable solution; if collisions instead are expected to be less frequent, a naked and flexible structure is aerodynamically and energetically more efficient.

In principle, these collision-resilient and fast drones that we are imagining could be applied in a number of situations, from search and rescue to surveillance and inspection. But the importance (and the challenges) of parcel delivery must not be underestimated. Having a drone deliver packages of various sizes and weights safely and reliably to humans still poses many challenges to researchers, especially if the drone has to figure out a route through buildings. And yet when we get there, we will remove at least one source of traffic from our roads: small-size and medium-size parcel delivery, which is currently transported with motorbikes, cars, and vans and is projected to increase exponentially as same-day delivery service becomes the de facto standard for commercial delivery. Some studies, such as one published in 2018 by Joshuah Stolaroff from the Lawrence Livermore National Laboratory in the United States, show that battery-powered drones could theoretically reduce greenhouse gas emissions caused by parcel delivery up to 54 percent compared to the current gasoline- and diesel-based deliveries (a lot depends, of course, on how the electricity that recharges the batteries is generated).[9] Assuming that self-driving electric cars will do their part and reduce passenger traffic, the cities of the future could indeed become more pedestrian- and cyclist-friendly, with more green areas and fewer traffic jams. Zipline, a Californian company, began delivering blood products to transfusion clinics in rural Rwanda using fixed-wing drones in late 2016 and today routinely flies a fleet of autonomous drones between their logistics centers and the clinics. Zipline's winged drones, roughly the size of a human adult, release their cargo with a parachute at the desired location and then fly back to the logistics center, where they are safely caught in flight by a rope hanging between two poles, a simple solution that does not require sophisticated perceptual intelligence and can operate also at low visibility. Matternet, another US company, instead uses quadcopters to operate between hospitals in urban environments where drones must vertically take off and land between buildings. Large companies such as DHL, Amazon, and JD already have experimented or are developing their own drone delivery systems.

Delivery drones tend to be larger than drones used for imaging and inspection because the lifting capability of a drone, that is, the weight that it can carry, is proportional to the surface of its propellers or wings. Large drones are dangerous when flying in proximity to people and require secure infrastructure to take off and land, to load and unload their cargo, and for storage and servicing when not in use. In other words, delivery drones need infrastructure akin to small airports. What would a drone airport look like? Sir Norman Foster, the inventor of modern airport design and architect of the largest airports in the world, explored the design of drone airports for deliveries in Africa. Droneports, as Foster called them, are organic, almost life-like structures assembled from arched modules made of natural materials available on site.[10] These droneports would serve not only as a logistic and technical hub for cargo drones, but also as a community center where people meet to exchange goods, receive medical care, upgrade their skills in robotics technologies under the remote supervision of instructors on other continents, and have access to rapid prototyping and computing facilities to develop their own businesses.

In the future, droneports may also pop up in industrial suburbs, on top of hospital buildings and on mall roofs, because large cargo drones, contrary to smaller personal drones, will need trained professionals for loading and unloading them. In order to democratize the use of delivery drones and let any member of the public and small shop owners rapidly exchange small items with the same ease and speed of exchanging text messages, drones must be mechanically safe when they approach people and must be sufficiently small to be stored away in a desk drawer or a backpack when not in use. To this end, in our lab we are rethinking delivery drones as a wrapping structure made of a folding mesh with integrated propellers that wraps around the parcel and serves not only as an air-lifting mechanism but also as a protective exoskeleton during flight in proximity to buildings and persons.[11] The drones look like flying balls that can be safely grabbed in flight: as the recipient opens the drone's mesh to grab the parcel inside, the propellers automatically turn off. These folding cargo drones shrink by almost 90 percent for convenient storage in a drawer or a backpack when not in use and can be transported just like a laptop. Larger versions of these delivery drones, similar to what we imagined in Jane's delivery, have extendable propellers that retract into densely meshed exoskeletons when they approach buildings and humans, akin to box turtles that pull in their legs when they are manipulated.[12] The dense mesh prevents even children's fingers from entering in contact with the large spinning propellers but greatly increases the use of energy required to stay in the air. Therefore, as soon as the drone is high in the sky and away from obstacles, it extends the propellers out of the cage and flies like any other conventional multicopter over long distances.

HONG KONG, 2037

Almost one year had passed since that sunny morning when Jane had welcomed the drone on the rooftop and with its help had given the finishing touches to her and Michael's project. A few months later, they had won the bid, and after a quick celebration (a bottle of champagne delivered at Michael's studio by—guess what?—another drone), they had started applying for permits and laying out the operational plan.

Now here they were, after some more frantic weeks of organization. Construction work was underway. Hard hat on her head, Jane was monitoring the traffic of workers, robots, and drones as the former epicenter of the city's car jams was being turned into a space for families with a culture center. Her assistant was up there with a team of workers, fixing the electrical line on the top of the six-story building that would become the public library, and got her on the walkie-talkie to ask for some backup. From the control station on the ground, she fixed three more drones with construction materials and sent them up there.

Drones helped transport tools and materials up and down, install wires and lighting, and secure external decorations. Sometimes two or three of them teamed up to transport long rods or heavy platforms. Other times they would deliver parts and tools to workers, mount and fix external parts, apply paint, check the stability of what human workers had just built, or simply clean up the finished parts.

Just like delivery drones had transformed the structure of cities, construction and inspection drones had turned the construction industry on its head. At first, they made inspection and maintenance safer and more efficient, taking upon themselves to routinely inspect buildings both outside and inside, as well as bridges, tunnels, power lines, and railways—resulting in fewer accidents for workers and less damage from fires or structural failures. Then, they went to work in indoor manufacturing, and from there they transitioned to outdoor construction sites. Now, they were making it possible to build a skyscraper at a pace of little more than four days per floor: it was the speed at which the Empire State Building was built in 1931, before more stringent building codes and safety rules for workers almost tripled the average time needed to build high-rises. Working conditions for workers became safer than they had ever been—thanks to drones that did the most dangerous parts of the job—but construction times were back to what they were in the early twentieth century. Watching her drones at work, Jane could not help noticing how they resembled birds of prey when they swoop, extend their arms to pick up an object, and then take off again without touching the ground; and when they reached high to deliver a tool to a worker, they sometime resembled parent birds that reach their nest to deliver food to their offspring. Indeed, inspiration from birds allowed researchers to progress from initial experiments in the 2010s to the drones that were now helping turn Jane's project into reality.

LONDON, UK, AND SEVILLE, SPAIN, CURRENT DAY

Despite being a bioinspiration enthusiast, always on the lookout for new sources of ideas for his robots, Mirko Kovac doesn't like the word *drone* that much. The problem, he has said in interviews, is that most people now associate it with the kind of UAVs (unpiloted aerial vehicles, the shortest and most technically appropriate name for flying robots) used by the military sector, whereas he is mostly focused on showing their peacetime applications.[13] He has repeatedly participated in the Drones for Good competition, held every year in the United Arab Emirates and focused on humanitarian and public service applications of flying robots, walking home with one of the top prizes in 2016.

But even in biological terms, he has a point: the word *drone* is somehow limiting. Male bees, the original drones, cannot really do that much, apart from flying and mating with their queen bee. All the practical work in a bee swarm—building the beehive, carrying around pollen, and making honey—is done by female worker bees. Kovac wants to give his flying robots the ability to do more interesting things than just flying around and watching from above, the typical function of today's surveillance and inspection drones.

Kovac, who divides his time between London's Imperial College and the Swiss Federal Laboratory for Materials Science and Technology, has spent the past few years combining two main research interests. The first one is multimodal mobility. "I want to build flying robots that can transition between water and air, or between air and a solid infrastructure," he explains. "Drones that can dive and then jump out of water and fly again, or that can perch on a pole, a tree branch, a wire and take flight again from there. These are all things that birds can do really well, so we obviously look at them for inspiration."

The second interest is manufacturing with drones. To that end, Kovac is merging UAVs with another headline-grabbing technology: 3D printing or additive manufacturing, the technique that promises to revolutionize manufacturing and allows building parts on-the-spot by depositing layer upon layer of melted materials that quickly solidify into the desired shape. The inspiration, Kovac explains, came from swiftlets—birds that build their nests, often in dark and dangerous caves, by depositing saliva. Though building nests with leaves or wood is quite common for birds, Kovac was fascinated by these "flying factories" that are able to carry inside them all that is necessary for building a home.

At his London lab, Kovac has been equipping quadcopters with a GPS—the easiest way to let a drone know where it is—and cartridges filled with chemicals that can be mixed on the spot and turn into a foam that can be sprayed on a surface, where it will quickly become solid. Kovac and its team scooped the first prize at the Drones for Good competition in Dubai in 2016 for demonstrating high-flying leak repair with a drone. At industrial sites, chemical leaks can be a threat to workers' health and can cause fires or damage to equipment. Leaks are often difficult to locate and reach, and the operation can be dangerous for humans. In such situations, drones are used in the best case as flying sensors, helping to locate and estimate the leak, but human operators need to intervene and fix it. "We showed that a flying vehicle can go to a leaking point on a gas pipeline and deposit material precisely to patch the leak," he explains. Kovac's Buildrone quadrotor can hover on the leaking pipeline, using an extended arm to achieve better balance and stability and precisely deposit a liquid polyurethane foam that expands and solidifies in just five minutes.[14] For the moment, the drones are controlled by a laptop that receives data from the onboard GPS and from infrared cameras that monitor the scene. One day, Kovac hopes, the drones will have their own cameras onboard and all the computing power to fly and work autonomously, possibly coordinating their work in swarms. These swiftlet-inspired robots could reach the top floors of high-rise buildings, as well as offshore platforms, power-line towers, and dams and carry out repairs that would be dangerous for humans. Or they could just as well work side by side with humans during construction work at great heights, making the jobs of humans easier and safer without displacing them entirely.

But before doing that, a drone—pardon, a flying robot—would need to stabilize or find something to attach to while working. The second option is indeed more interesting because it would allow the robot to save power, and battery life is always going to be a precious resource for drones. Once again in search of inspiration from animals, Kovac looked at how spiders cast nets that support them while they wait for their prey. Kovac's flying robot called, not coincidentally, the SpiderMAV, has two modules filled with compressed gas, that can each launch a polystyrene thread with a magnet at its end.[15] The thread is launched, Spider-Man style, toward a metallic surface onto which the magnet can stick, such as a beam. Then a spool reels the thread until it has enough tension. One module, on top of the robot, does the main job of perching, while a second one, shooting its thread from the bottom, acts as a stabilizer. Once the two threads are stretched enough, the drone can slow down or even stop its propellers and is stable enough to turn into a flying 3D printer even when working outside and facing winds.

Kovac is not the only scientist who is trying to expand drones' repertoire with manipulation and perching skills. At the University of Seville, in Spain, Anibal Ollero has several years of research under his belt on aerial manipulation. In the mid-2010s, working with several other European researchers in the ARCAS project, he achieved the first demonstration of how a group of autonomous flying vehicles can cooperate in inspection and construction tasks.[16] Ollero and his colleagues fitted robotic arms on quadrotors and on small, autopiloted helicopters, and studied the problem of how to simultaneously control the vehicle's flight and the operation of the arm. This is easier said than done. There are tried-and-tested control strategies that can do one of the two things at the time, so one could naively think that coupling a standard autopilot to a standard manipulator controller (such as the ones used for industrial robotic arms) would do the job. But in fact, the moment a moving arm is attached to a flying vehicle, it starts altering the flight dynamics, even more so when it interacts with the environment. The movement of the manipulator changes the helicopter orientation and position; the arm's position then needs readjusting in order to reach the target object, which in turn changes again the helicopter's position, and so on. Ollero and his colleagues solved the problem by designing a new control strategy where information from the arm's force and torque sensors is fed directly into the software controlling the helicopters' rotor blades. At the same time, they taught the arm controller to use flight itself for achieving the desired position: when the object is within reach of the manipulator, it is often more convenient to move the whole rotorcraft toward it rather than extending the arm. In technical jargon, what Ollero did was to turn the aircraft's movement around the yaw axis (the movement a drone does when turning left or right while hovering) into an additional degree of freedom that the manipulator can use. These solutions allowed the ARCAS team to demonstrate in 2014 how a group of small, unpiloted helicopters fitted with industrial robotic arms (similar to the ones used in automotive manufacturing) can cooperate to pick up and transport large and heavy structures—for example, a platform that is transported to the top of a building on fire, allowing rescuers and survivors to transfer to the adjacent building.[17]

In another European project, AEROARMS, Ollero wanted to develop solutions for sending drones to do outdoor inspection and maintenance at great heights—for example, in industrial plants, on power lines, or on high-rise buildings. A whole family of flying manipulators resulted from that project, including an hexarotor with two articulated grasping arms (it is indeed hard to do any manipulation task other than simple grasping or pushing with just one arm) and an eight-rotor drone fitted with an arm carrying an ultrasound sensor on its tip. The latter proved capable of navigating through a dense network of pipes, flying along and around one pipe and maintaining the sensor in contact with it while avoiding collision with other pipes above and below.[18]

"Inspection and maintenance were the first industrial applications of flying manipulators," Ollero points out. "Next will come aerial coworkers in indoor manufacturing and also in outdoor maintenance, that can interact with people in a safe way; for example, drones providing small parts or holding objects for a human working several meters up performing riveting tasks. Having many servicing drones flying outdoors and delivering packages will take more time."

Ollero is now busy designing morphing drones that combine the respective advantages of fixed-wing vehicles (long range and energy efficiency) and rotary wings (the possibility of hovering and moving in narrow spaces), as well as manipulation: drones, in other words, that can take off and land vertically like helicopters, then morph into fixed-wing vehicles to fly at high speed and for long distances, then morph again into rotorcrafts to slow down and approach the working site, and pull out arms and hands for manipulation. This frequent switch between different flight modes and between flight and manipulation would allow drones to conduct long-range missions such as inspecting power lines over a large area. And it is exactly the sort of multitasking that many birds effortlessly accomplish.

"Birds are able to fish while flying, or to provide food to the small birds in the nest," he notes. "They do have manipulation abilities that require maintaining equilibrium at all times. These are very, very difficult things to do with drones without significant research and development. Planning strategies are another problem. Look at how birds manage to land on a wire. It requires significant planning abilities that right now we don't know how to implement."

A challenge for the future will be to add softness to aerial manipulators as well. "It is necessary in order to have robots that can interact more safely with humans, and at the same time have full manipulation capabilities—and I stress full, not limited to one particular application. If you combine this ambition with the ambition of being soft and to reduce energy consumption, you realize we still need many years of research," Ollero concludes.

If Kovac, Ollero, and the other scientists working on aerial manipulation have their way, such drones will initially find niche applications in disaster management or infrastructure maintenance in sparsely populated areas—just as autonomous driving will have to be tried in mines, factory lots, and corn fields before hitting the road. But if the technology gets to the point where drones can reliably perform a range of manipulation tasks while interacting safely with people, it will be only a matter of regulation catching up, for aerial manipulation to become part of the scenery in most cities.

HONG KONG, 2037

The city was almost dark when Jane finally decided to leave the construction site. Most workers had gone home a couple of hours earlier, when their shift ended, but drones could still keep working after the sunset thanks to their sensitive cameras and some help from nonvisual sensors such as GPS, so their shift lasted a little longer. But Jane needed to stop. It had been a long day, and the following day was going to be even longer. A few keystrokes on her laptop were enough to send all the drones to sleep in their shelter near the construction site, which she proceeded to lock digitally. She then pulled up the taxi app on her phone and summoned the vehicle that would take her home.

The taxi arrived less than a minute later. As it swooped toward the construction site on its fixed wings, then pulled out rotors to hover down and land, it was obvious to Jane that she was watching a larger-scale version of the same technologies that she had witnessed when the delivery drone had reached the droneport on top of her office's building—the same technologies she had watched all day at work, allowing manipulator and coworker drones to operate above and around the construction site. Just like those drones, this flying personal transportation vehicle used vision-based algorithms and collision-avoidance systems to fly at high speeds above the city, shuttling people instead of packages around. It was still a relatively new service in the city, and many people still referred to these vehicles as "flying cars"—a term that the popular press has used for decades, since the twentieth century, to evoke the vision of personal transportation vehicles with wings instead of wheels. Now this vision was finally true at a time when cars themselves (mostly autonomous, hyperconnected, shared rather than owned) had less and less in common with their twentieth-century predecessors.

Most people in the city still preferred to move on wheels—on autonomous ground taxis rather than flying ones. It was understandable. If you fear flying, as many people do, having a neural network at the control instead of a human pilot does not make it better. But Jane had seen autonomous flight at work for so long now and in so many different situations that she'd come to trust it entirely. She jumped into the taxi, relaxed on the reclining seat, and once again marveled at the sight of the city from above while the drone zipped over it and toward her home.

5 LOVE AND SEX WITH ROBOTS

Exchanging rings was the most difficult part. For all the remarkable skills April had that convinced Jack to marry her, picking that tiny white gold band from the crimson cushion, holding it between her two fingers, and gently slipping it on Jack's ring finger was a bit too much for her. They'd rehearsed it a couple of times, but the ring kept dropping. "I think I'm too nervous," April had commented the second time, as they looked for the ring under the sofa where it had rolled, and decided to let it go. There was going to be enough nervousness in the room without guests having to crawl on the floor to look for the ring. So after the solemn—and world-first—yeses were uttered, Jack helped April put on her ring, then put the other one on his own finger.

"It's sweet," Jack's mother whispered to her husband. "Still strange, but sweet." Jack's parents were sitting a few steps away from the newlyweds, in the small ceremonial room that the city hall devoted to weddings. They were surrounded by a few of Jack's brothers and his sister and a few of his closest friends. Toby and Frances, the kids from his previous marriage, were holding flowers and beaming with joy. Helen Myzaki, the engineer who considered April her "most successful protégé," was also there, at Jack's insistence.

In front of them was the perfect marriage scene: both groom and bride were smiling and visibly emotional. Jack looked happier than he had been in years. April looked overwhelmed with emotion—Myzaki had made sure she would—and stunningly beautiful. Then the kiss came, greeted by a short applause.

"I will never get used to it, I guess," Jack's father sighed, his voice drowned out by applause. "Let's hope the people out there do. Let's just hope they leave them alone."

April and Jack had gone to great lengths to avoid publicity and keep the date secret, but somehow a leak had reached the media. A small band of reporters and curious people armed with cameras were waiting outside, ready to live-stream the moment when April and Jack would come out of city hall and stroll toward their car to go on their honeymoon. There was no way around it: the first human–robot marriage was going to be prime-time news, no matter what they did to avoid it.

One week before the wedding, Jack had agreed to do an interview with a local journalist who had a large social media following. He was hoping the interview would at least present his side of the story and would stop fake news from spreading.

In the interview, viewed about 1 million times in the week leading up to the wedding, he recalled how it had all started. After his wife's death, he was left with two children to care for, and he needed to work two jobs to make ends meet. Dating was not an option: he had no time. So he had resorted to other options for sex. He tried a few—from a wildly popular virtual reality pornography service to dolls of varying skin and hair color—before settling on April, a sex robot marketed by Myzaki's Tokyo-based Soulmate company that was a total novelty at the time. In fact, he'd received Soulmate's third robot. At the end of the three-month leasing period, he decided to upgrade to what Soulmate called the "long-term relationship" model: buying the robot.

The other sex robots he'd tried could do the job, but April had turned out to be something else. Her soft and sensitive artificial skin could transmit tactile sensations and react to touch more realistically than all her predecessors. Her movements, governed in part by old-fashioned electric motors as well by pneumatic actuators in strategic places, could almost match the pace and agility of human movements. And her neural networks could use all sorts of clues—what he did when they were together, as well as his online searches—to learn about Jack's personality, mood, and fantasies.

He quickly developed a bond with April—or, as the people who knew about it preferred to say at the time, he became addicted to her. Sex had a lot to do with it, but at some point, he realized he just needed to have her around. In the morning, she would sit at the kitchen beside him, and they would exchange comments on the day's news. On summer evenings, she would ride with him along the river on their bicycles as they shared thoughts on the beauty of nature and the future of humanity. And at night, she would be at his side on the sofa after dinner, watching his favorite TV shows with him, and commenting on characters and plot twists. For him, those casual discussions brought back the feeling of being in a relationship—no matter that, as he knew very well, the only reason she could comment on TV shows was that she was browsing discussions on social networks about them while sitting on the sofa, and using her neural logic inference engine to build sentences from the most popular and contrasting opinions.

Things became more complicated when his kids became friends with her too. At first, he'd told them she was there to help with house chores; she did some of them in fact, though it was not really her strong suit. But his elder daughter, Helen, way smarter than the average seven-year-old, had probably figured out herself before he sat down with her and nervously tried his best to explain what was going on. In fact, the kids had seen their dad lighten up again for the first time in years, and they ended up loving April's presence in the house—even more so because she was clearly not their mother's replacement. How could she be?

Then came Jack's car accident. With multiple injuries and a suspected head trauma, he had to spend three weeks in the hospital. Being separated from April proved unbearable, yet he could not convince the hospital staff to let her in his room. The official answer was that she was not certified for a clinic environment. Plus, he had to ask his parents to take care of his kids and the house full time, when in fact there were simple things—checking on them while they did their homework, feeding the dog, preparing breakfast—that April could easily do, if only the law allowed the kids to be alone with her. Also, he realized he was going to need April more and more as old age approached and that her being considered just another appliance by the law was going to be a problem.

He began to search online for options and one night stumbled on an old interview of David Levy, a businessman, author, and artificial intelligence expert who in 2017 predicted that a human would marry a robot by 2050. Levy's work provided Jack with some basis to begin a long legal battle that—although it made him a laughingstock to many people in his home country—finally gained some traction in Massachusetts, where he ended up moving with the kids and April. The rest was recent history: the announcement, the planning of a small party that he would rather avoid but that the kids insisted on having, and today's ceremony. In the end, Levy's prediction had been fulfilled only a few years later.

CALIFORNIA, JAPAN, AND EVERYWHERE, CURRENT DAY

A common question that roboticists often ask themselves or—even more often—are asked by tech journalists, is: What will be the killer app for robots? What application will finally do the trick of turning robots from the highly successful but niche product they are today into a mass-marketed consumer product, finally fulfilling the prophecy (made among others by Bill Gates in a cover article in *Scientific American* in 2008) of "a robot in every home"?[1]

Coming up with a convincing answer is not obvious. The robotics market today is mostly about industrial and service robots, expensive and often sizable machines bought in relatively small numbers by corporations rather than individuals. As for all the experimental work going on in research labs, a lot of it revolves around preparing robots for quite extreme situations: search and rescue during disasters, assistance to the disabled, exploration of hostile environments. However important they may be and as many lives they may save or make better, such robots could never sell in millions.

To date, the only mass-market robots are autonomous cleaning robots, among which Roomba, one of the very first vacuum cleaning robots by the US company iRobot, has sold almost 30 million units since 2002.[2] At the time, many were surprised that iRobot had decided to enter the market for vacuum cleaners. It had hitherto focused on military robots, and its founder was former MIT professor Rodney Brooks, one of the godfathers of modern robotics science. That such a company and such a scientist were now working on vacuum cleaners seemed a waste of talent. It turned out that Brooks was right: applying some state-of-the-art sensing and navigation technologies to a domestic appliance provided a fast way to get robotics into many households.

When it comes to ways future robots could make themselves useful in future homes, though, there may be more entertaining options than vacuum cleaning. If history gives any guidance, the killer application that drives the adoption of a new technology often has to do with sex. The invention of the printing press is most often credited with allowing the birth of modern journalism or a more liberal circulation and discussion of the Bible, but in the sixteenth century, quasi-pornographic works such as François Rabelais's *Gargantua and Pantagruel* had a big role in popularizing printed works and creating a market for them. The vibrator was one of the first electrical devices to be built and sold; it was patented in the 1880s and initially used as a medical device to cure women suffering from "hysteria."[3] Pornography has sold millions of papers and for decades has been one of the easiest shortcuts to attract readers at the newsstand. Porn was instrumental in the adoption of home video and one of the main reasons people began to go online at the dawn of the Internet era.[4]

In all those cases, the market share of sex and pornography declined over time as the technology behind each new media became mainstream and the market grew. But sex was a key driver for creating an initial market that justified investments and started economies of scale. Sex, in other words, often drove early adoption of new technologies. It makes a lot of sense to expect that sex will somehow contribute to plant the first seeds of a mass market for home robotics too. Of all the adventurous predictions contained in this book, this may in fact be the safest bet.

So "let's talk about sex robots," as the scientific journal *Nature* encouraged society to do in an editorial in 2017, acknowledging that "technological developments in soft robotics and artificial intelligence put these machines on the horizon, at least in basic form."[5]

Technically, sex robots are already on the market. They are upgraded version of sex dolls—not the inflatable ones, but the high-end lifelike versions, made out of silicone or elastic polymers derived from the special effects industry, with realistic hair and articulated skeleton and joints. In 2018, the California-based company Abyss Creations—a successful manufacturer of sex dolls for quite some time—launched Harmony, essentially a combination of a silicone sex doll; a robotic head with moving neck, jaw, mouth, and eyes; and an AI-powered app that allows the robot to speak with a Siri-like realistic voice. Through casual conversation (What is your favorite food? How many brothers and sisters do you have? What's your favorite color?), Harmony learns about her mate and stores information that may be used later in conversations. The skin is heated to match the temperature of the human body, and sensors on the highly realistic skin allow the robot to react—moving, speaking, moaning—when touched. The whole system costs around twenty thousand dollars and can be added to a sex doll of choice, which in turn ships for a few thousand dollars. A similar technology is behind the "robot companions" that Green Earth Robotics manufactures in Canada.

In a 2018 segment about Harmony on *ABC News*, journalist Katie Couric visited Abyss's premises and interviewed an early adopter of Harmony who was stepping by to check out the finishing touches to his purchase. The man—not showing his face, but probably in his fifties or early sixties judging from his voice—was obviously impatient to bring Harmony home and "physically interact with this piece of art," as he put it. He described how, despite having previously been married for fifteen years, he was confident he would not miss the intimacy and deep relationship a real human being would provide, instead appreciating how Harmony would never lie, never cheat, always be honest. Asked by Couric whether he worries about other people finding all this "super-weird," he replied that "in 20 years it'll be normal. I think everybody will have some form of robotic companionship."[6]

If you saw *Lars and the Real Girl*, the 2007 movie in which Ryan Gosling's reclusive character develops an intimate relationship—in fact, his first intimate and caring relationship—with a sex doll he's ordered online, you know how disturbing and yet not totally unimaginable this perspective may seem. And that doll could not move, speak, or even look Ryan in the eye. Now think of crossing that movie with the equally disturbing—but even more relatable—*Her*, where Joaquin Phoenix's character falls desperately in love with his vocal assistant (voiced by Scarlett Johansson), only to discover that she's parallel-computing relationships, having thousands of similar ones with other users at the same time. To think that people would develop affection, attachment, and even something resembling love for a realistic, sexually attractive, and satisfactory humanoid with a warm voice and the ability of having conversation and remembering about your birthday no longer sounds that far-fetched, provided, of course, that the technology works.

Once you subtract the hype, existing sex robots still have a long way to go before being able to provide a realistic experience. After all, most humanoid robots can barely walk a few steps without falling, move their arms and legs at a frustratingly slow pace, and have a hard time opening a door. Unless you have very low expectations, sex requires a level of agility and a refined control of limbs, head, and joints that robotics will struggle for some decades to achieve. As impressive as Boston Dynamics' back-flipping robots are, covering them in a warm silicone skin and adding a vocal assistant falls short of being a sexual fantasy realized.

One problem is that manufactured actuators (a combination of a motor and a gear) do not have yet the efficiency, compliance, and flexibility of biological muscles. Electric motors can be more powerful than biological muscles but come at comparatively higher weight per unit of power and are rigid. Pneumatic actuators that rely on compressed air or fluid, such as those found in some construction drillers and robots, can be more powerful too, but they require relatively bulky and noisy compressors. Pneumatic artificial muscles, first described in 1957 by atomic bomb engineer Joseph Laws McKibben, are inflatable tubes constrained by a stretchable mesh that makes them bulge and shorten when pressurized. While this technology can effectively mimic the behavior of biological muscles, it still requires an external compressor. However, the soft robotics community is making good progress at addressing some of these challenges. For example, a team of Swiss and Japanese researchers has shown a fiber-like, soft pump made of a thin elastic stripe filled with saline water that moves along the stripe length when a small current is applied.[7] These soft pumps could be integrated with pneumatic artificial muscles and operate as silent, lightweight, and flexible power generators[8]. Researchers are also investigating a variety of soft elastic membranes, such as dielectric elastomers and shape memory polymers, that rapidly deform in the presence of small electrical currents. Although these technologies cannot apply large forces, their light weight, softness, and manufacturability in diverse shapes makes them promising for artificial muscles that do not require much strength, such as those involved in facial expressions or other tiny parts of the body.[9]

Another problem is that independently of the actuation system, the agility, flexibility, and strength of human bodies cannot yet be matched by humanoid robots, which are mostly made of rigid parts designed and assembled after classic engineering principles of electromechanical efficiency. Instead, human bodies are made of an interconnected network of bones and muscles, also called a musculoskeletal structure, that can deform in hundreds of directions, absorb and release energy like a spring, and rapidly shift from a soft passive mode to high strength. Recent musculoskeletal machines represent a promising research direction to approximate those capabilities. They are made of rigid parts that play the role of bones, linked by joints and at least a pair of tendons that operate in agonistic-antagonistic mode, similar to the structure of a biological vertebrate. When both tendons are relaxed (that is, not under tension), the articulation can be passively moved around its joint; instead, when both tendons are under tension, the articulation is stiff and requires high external forces to be moved. By selectively applying more tension to one of the two agonistic-antagonistic tendons, the articulation can be moved to one direction or the other while also controlling its stiffness by means of the overall tendon tension. While humanoid robots made of rigid parts, such as Asimo by the Honda corporation, have approximately 30 degrees of freedom (the number of directions in which robot parts can move), musculoskeletal robots have approximately 60 degrees of freedom. They are more agile but still a far cry from humans, which have more than 400 degrees of freedom. Enter Kengoro, a musculoskeletal robot designed at the University of Tokyo to mimic the proportions, weight, and skeletal and muscular structure of the average human.[10] Kengoro has 114 degrees of freedom (27 percent of the human body), and when the hands are included, it reaches 174 degrees of freedom. Kengoro's muscles are composed of an electrical motor, mechanical parts, a wire, and sensors and are attached to the robot's rigid links (bones). The robot, which, contrary to other humanoids, has been designed to leverage physical contact of its entire body with the environment, faithfully reproduces 51 percent of the muscles in the human body. In order to dissipate the heat generated by the large numbers of electrical motors that pull the tendons, the robot is equipped with a liquid cooling system that circulates through its body. The fascinating or freaky result, as *Wired* magazine described it, of this cooling system is that the robot leaks tiny drops of liquid under strenuous effort, for example, when performing push-ups, just like human bodies.[11] While the team at the University of Tokyo foresees a possible use of their robot in medical schools and in research as a model of the human body in motion, musculoskeletal technologies could one day be adopted for humanoid companions too.

The emotional aspects of creating intimacy with a robot may be even more challenging than the physical ones. Manufacturing androids with a highly realistic human appearance is not so difficult—visual effects specialists in the movie industry have been doing it for decades—as long as they stand still and are silent. But making them move and talk in a way that enables a natural, credible, and satisfactory interaction with people is a different story, and we are not yet quite there.

There are, however, two aspects of human psychology that could make our fictional story come true in a foreseeable future. The first aspect is that humans have an innate tendency to attribute complex emotional and social abilities even to simple animate objects. In the early 1950s, Grey Walter, a British neuroscientist and electrical engineer, built a series of mobile robots to show that complex, purpose-driven behaviors do not require complex brains.[12] As his shoebox-sized wheeled machines, equipped with a light sensor, bumpers, electrical switches, and hard-wired resistors, roamed through his living room, turning away from furniture and avoiding light until they reached a dark spot where they stood motionless, a BBC commentator depicted the robot's movements with emotional connotations typically used for describing animal behavior. Some thirty years later, Valentino Braitenberg, a neuroanatomist and former director of the Max Planck Institute of Biological Cybernetics in Tübingen, published a small book describing a series of imaginary vehicles with simple wirings of sensors and motors inspired by the anatomical and physiological design of nervous systems, such as symmetrical architectures, cross-lateral connections between left and right brain regions, time-delayed activity, and nonlinear transformation of input signals.[13] When placed in an environment, Braitenberg's vehicles displayed a range of complex behaviors that an observer might have labeled as aggression, love, fear, logic, foresight, and even free will. Although these imaginary vehicles inspired several roboticists and contributed to the birth of biologically inspired robotics in the last decades of the twentieth century, Braitenberg's goal was to show that much of the complexity observed in animal behavior stems from the interaction with the environment rather than from the complexity of the brain as neuroanatomists used to think.

The second aspect of human psychology that may facilitate social interaction and empathy with robots is that communication is a bidirectional process where both the signaler and the receiver play an active role. Cynthia Breazeal, a professor and associate director of the MIT Media Lab, made this point very clear almost twenty years ago by means of Kismet, a robotic talking head with moving eyes, eyelids, ears, lips, neck, video cameras, microphones, and a loudspeaker.[14] Kismet's behavior was driven by a set of stimulus-reaction rules triggered by external sensory signals. For example, the robotic head could direct its attention to a shared reference with a human, could recognize emotional states from voice features and give expressive feedback, and could take a proactive role in regulating the interaction with a human. Although the robot did not have a full understanding of human language, that did not seem to bother people, who engaged in long conversations and adapted to the robot's behavior, for example, by speaking more slowly, waiting longer for a response, and checking for cues from the robot. As Breazeal puts it, "Social interaction is not just a scheduled exchange of content, it is a fluid dance between participants. In short, to offer a high quality (i.e., compelling and engaging) interaction with humans, it is important that the robot not only do the right thing, but also at the right time and in the right manner."[15] Measures of human-robot coordination, also known as collaborative fluency, showed that higher reaction times translate into higher subjective perception of robot's fluency.[16] Embodied communication with robots is so strong that it can affect and modify human behavior. For example, a small humanoid robot developed by Catalia Health has been shown to engage and maintain people in complex treatment regimens, thus reducing supervision by a nurse or doctor. In her book *The New Breed*, MIT researcher Kate Darling offers plenty of evidence that robots with lifelike features and movement can elicit strong empathy, akin to the empathy that our companion animals elicit in us.[17] People will immediately put a robot down and feel guilty if the robot complains and moves helplessly. As Darling reports, life-like appearance is not even necessary to establish strong bonds between a human and a robot: owners of vacuum cleaning robots asked the manufacturer to repair their robot instead of sending a replacement, and soldiers working alongside demining robots risked their life to bring a robot back to safety from enemy fire. Today's voice assistants, such as Siri and Alexa, are very good at speech recognition, but their disembodied nature does not promote sustained conversations, not to speak of emphatic interactions, with humans.

An intimate relationship, however, and that is not only about sex, requires more than speech and gesturing. Affective relationships most often involve the feelings of touching and being touched, the warmth and texture of the other person's skin, the pressure, speed, and motion of a stroke. In many situations, these tactile sensations can convey information and feelings much better than words. For example, researchers showed that touching less accessible regions of a humanoid robot's body (say, its buttocks and genitals) is more physiologically arousing than touching more accessible regions (its hands and feet), suggesting that people treat touching body parts as an act of closeness that does not require a human recipient.[18] It is also important for robot companions to be able to perceive tactile feedback, which scientists call haptic information (from the Greek *haptesthai*, meaning "to touch"). Paro, a social robot that looks like a baby harp seal, leverages relatively simple haptic sensors under soft fur to understand how it is hugged, stroked, or patted and reacts with behaviors that induce positive feedback from a person. The robot, conceived by Japanese roboticist Takanori Shibata, has been deployed in hospitals and retirement homes since the early 2000s and has been shown to improve the mood of elderly persons,[19] help with depression,[20] alleviate behavioral disorders associated with dementia,[21] and even facilitate interactions among human patients and caregivers.[22] Skin is the largest human sensor and is finely innervated to provide us with a large variety of precisely localized sensations and help us in coordinating daily tasks in ways that we often don't even realize. Just try to pick up and light a match stick with cold-numb fingers (if you don't want to numb your fingers, you can find online videos of a scientific experiment whose subjects' fingers were temporarily anesthetized).[23] That is the challenge that robots today face all the time because they do not have sensorized skins and must rely uniquely on vision or other comparatively rudimentary sensors, and it is one of the reasons that robots are not as dexterous as humans at manipulation and other physical interactions.

In 2019, Gordon Cheng, a robotics professor at Technical University Munich, unveiled a robotic skin that can be wrapped around robotic bodies.[24] The robotic skin is a foldable surface made of hexagonal tiles that can perceive contact, acceleration, proximity, and temperature. Although the tiles are relatively large (one inch in diameter), they could help robots better understand their surroundings and support safer operation in the proximity of humans. While Cheng's team works on the miniaturization of his skin tiles, other researchers bet on soft and stretchable elastomers. For example, an MIT team directed by Professor Daniela Rus developed artificial skins made of conductive silicone membranes patterned after kirigami, a technique similar to origami that uses small cuts, in addition to simple folding, to enable large structural deformations.[25] These skins can be attached to any soft body to perceive deformations caused by external forces or by internally generated motion with the help of neural deep learning artificial intelligence, although we are still far from even approximating the variety of sensations, density of sensors, and conformability of biological skins.

Technology is only part of the story. Robotic companions will also need more sophisticated haptic intelligence to physically engage in a satisfactory relationship with a human. Hugging a person, for example, is a common behavior and can reduce stress and support empathy. A team directed by Katherine Kuchenbecker at the Max Planck Institute for Intelligent Systems in Stuttgart developed HuggieBot, a modified version of a commercial humanoid robot, to study the most important factors in hugging a person.[26] They showed that humans not only prefer to be hugged by a soft and warm body, but also prefer robots that can match the pressure and duration of their hug; in other words, they like to be moderately squeezed, and they expect the robot to rapidly disengage as soon as they release their arms. Furthermore, the team also reported that people who engaged in satisfactory hugs ended up with a more positive view of robots. All this bodes well for our story. As robots acquire more compliant, agile, and soft bodies, increase their perceptual and motor capabilities, and learn to adapt their social and physical interactions, humans may attribute more logic and emotional intelligence to robots and engage in more sustained and satisfactory relationships.

And what about those robot looks? What happens when a humanoid robot looks like a real person? Hiroshi Ishiguro is a professor at the University of Osaka who has devoted most of his career to building realistic androids (both male and female), having them interact with people, and conducting experiments to study at what point, and in which conditions, the feeling of being in company of a real person sets in.

One of his best-known creations is the series of Geminoid robots, developed over the course of several years between the late 2000s and the early 2010s[27]. The Geminoid is nothing less than a robotic duplicate of Ishiguro himself. Its exterior is made out of silicone skin, molded by a cast taken from its creator's face, and hand-painted to render face details and textures. The robot has about fifty pneumatic actuators, allowing it to move its face, arms, and torso smoothly—albeit slowly. The Geminoid cannot walk; it was designed to have conversations with people around a table while being remotely controlled by a human operator—ideally, the very human original after which it is modeled. Ishiguro devised a remote control interface that captured his own lip movements by an infrared motion-capture system and turned them into instructions for the pneumatic motors that move the robot's lips, sent over a data link together with Ishiguro's voice. This way, the human operator could just talk, while the robot would reproduce all aspects of his speech (sounds and lip movements) at another location. Meanwhile, Ishiguro could monitor the robot's surroundings with cameras and microphones and could use a graphic interface to easily control the rest of the Geminoid's 50 degrees of freedom with a single mouse click. The overall design and control of the robot were focused on the body parts and movements that allow dialogue interaction: the robot cannot manipulate objects, but it can nod, shake his head in disagreement, smile, and stare.

The Geminoid project gained Ishiguro fame and quite some media coverage, including several photo opportunities where he appears alongside his robotic double; in still photos from that period, it is not always easy to tell the copy from the original, although the difference becomes obvious in videos. But Ishiguro's main research interest was to study how people would react and interact with it. Would people interacting with Ishiguro's double undergo the same suspension of disbelief that sets in when we watch a movie and empathize with the characters as if they were all real? Would they start to feel as if they were in the company of the real human?

Anecdotally, Ishiguro reported that most people, upon first meeting with the Geminoid, began to have "weird and nervous feelings" as soon as they realized it was a robot, but then found themselves concentrating on the interaction, to the point that the strange feelings soon vanished. In fact, in his experiments, Ishiguro discovered a few interesting things not only about people's reactions but about his own as well. While operating the robot remotely, he found himself "unconsciously adapting movements to the Geminoid movements. . . . I felt that, not just the Geminoid but my own body is restricted to the movements that the robot can make." In other words, he quickly developed a feeling of embodiment.

A lot has happened in robotics since the early 2010s—in fact, most of the technologies we describe in this book emerged after the Geminoid project—and Ishiguro has more recently taken advantage of advancements in soft robotics and artificial intelligence, particularly the AI branch that deals with natural language processing, to build Erica, a female autonomous android that can engage in conversations with people. Unlike Geminoids, it does not need remote control—and like them, it has quickly become a media sensation in Japan and beyond. Erica's actuation (based on pneumatic systems) and degrees of freedom are similar to her predecessor, but AI allows her to process language in real time, giving more-or-less appropriate responses to casual conversation. For example, she asks, "Where do you come from?," and when the person answers "Kyoto," she comments, "Ah, not far from here." She also uses vision and image-recognition algorithms (similar to those that power Google's image search) to detect the presence of people in the room, follow their gaze and engage with them, and do some basic reading of facial expressions and, thus, emotions. According to the *Hollywood Reporter*, at the time of writing this book, Erica is working as an actor in a movie production where the robot plays herself in the story of a scientist and his android creation.

Erica is not alone. Sophia is a female android with a realistic face and expressions that relies on artificial intelligence to engage in communication with humans. Sophia too has played herself in an episode of the TV series *Westworld*. Sophia is produced by Hanson Robotics, a Hong Kong company founded by roboticist David Hanson, who sees his android creations as platforms for research in artificial intelligence, for medical studies, and for entertainment. Unlike Erica, Sophia does not attempt to conceal her robotic nature: the back of her skull is transparent and reveals the motors that pull the muscles and tendons of the face, and her body is noticeably robotic. There may be a reason for this identity disclosure. More than fifty years ago, Masahiro Mori, then a robotics professor at the Tokyo Institute of Technology, wrote an essay on the quality of feelings elicited by robots.[28] In Mori's view, feelings become more positive as robots become more human-like, but there is a point along that progression where humans start to experience uncanny, eerie feelings; however, feelings become positive once again as the robot's progression toward human resemblance approaches 100 percent fidelity. Mori coined the term *uncanny valley* to indicate the small region of uncanny feelings surrounded by two peaks of positive feelings. Although uncanny feelings have been anecdotally reported when humans are exposed to realistic androids, it is still debated when and what elicit those feelings. Some researchers argue that they arise when humans discover that what they previously thought was alive is in fact a machine; other researchers argue that they arise when parts of the brain detect a mismatch between high-fidelity features, such as visual resemblance, and less realistic features, such as slow motion.[29] Some researchers think that the uncanny valley is not a valley but a cliff from where there is only a way down, whereas other researchers think that it is not a problem because humans develop different feelings depending on their expectations of and interactions with those robots. Whatever the case, a robot companion that manifests its true nature by its looks or behavior may be a safer bet to reassure and induce humans in behaviors that facilitate sustained interactions.

According to the British researcher Kate Devlin, the best sex robots may not look like humans at all, but could apply soft robotics technologies to broaden the scope of human sexuality. For two years, between 2016 and 2017, Devlin has organized a "sex tech hack" at Goldsmith University in London, challenging its students' creativity on the topic of how robotics technology can apply to sex. As the *Guardian* reported, one of the most impressive—and sexiest, in Devlin's own assessment—ideas was a bed equipped with inflatable tubes made out of plastic. The user would lie down and be hugged, squeezed, touched, enveloped by the tubes, pulsating with air.[30]

Online pornography is often accused of generating addiction and contribution to sexual and emotional disorders—although, as it happens whenever the words *addiction* and *online* come together, solid evidence is hard to come by. If sex robots become a superrealistic and more satisfactory form of pornography, it may indeed make some of its users reclusive and emotionally unbalanced.

Those who see the glass as half full, though, note how they could provide sexual relief and—yes—some form of companionship to elderly people, disabled patients, possibly prisoners in jails.

David Levy, the scholar who wrote one of the first books on the subject in 2007 and formulated that prediction about the first human-robot marriage by the mid-twenty-first century, is overall optimistic. He notes that sex robots could replace people—mostly, though not exclusively, women—in the oldest and possibly least desirable occupation on Earth, and he sees that as a good thing.

"I started analyzing the psychology of clients of prostitutes" he told *Scientific American* in a 2008 interview.[31] "One of the most common reasons people pay for sex was that people wanted variety in sex partners. And with robots, you could have a blonde robot today or a brunette or a redhead. Or people want different sexual experiences. Or they don't want to commit to a relationship, but just to have a sexual relationship bound in time. All those reasons that people want to have sex with prostitutes could also apply to sex with robots."

Levy's view is pragmatic. Prostitution has existed since the dawn of time, and nothing—not laws or education—seems to be able to make it go away. Those who buy sex objectify the seller, and that means that buying sex from a machine—provided it is realistic enough—would not make much difference. If it works, this is one of the cases where replacing human workers with machines could make everyone happy. Maybe not that many people will actually buy a sex robot for their home, but robotic brothels could become a reality—once technology catches up—and one in many ways preferable to the current reality of prostitution.

"I don't think the advent of emotional and sexual relationships with robots will end or damage human–human relationships" Levy continued. "People will still love people and have sex with people. But I think there are people who feel a void in their emotional and sex lives for any number of reasons who could benefit from robots. Other people might try out a relationship with a robot out of curiosity or be fascinated by what's written in the media. And there are always people who want to keep up with the neighbors."

Levy's arguments do not convince Kathleen Richardson, a professor of ethics and culture of robots and AI at De Montfort University in Leicester in the United Kingdom. In 2015, she launched the Campaign Against Sex Robots, in effect proposing a ban on the development and sale of all Harmonies present and future. "We think that the creation of such robots will contribute to detrimental relationships between men and women, adults and children, men and men and women and women," she told the BBC. She warns that sex robots—humanoid ones modeled on women, at least—would be effectively artificial women slaves, further encouraging the male-centric perspective on sex, power, and ownership of women's bodies that is the substrate for discrimination, violence against women, and rape. "The dolls are modeled primarily on pornographic representations of women," she wrote in an article in the *IEEE Technology and Society* magazine in 2016.[32] "The speech programs in these robots are primarily focused on the buyer/owner of the model. While the pornography industry shapes the way the robots are designed in appearance (their ethnicity and age are also important), the type of relationship that is used as the model for the buyer/owner of the sex robot and the robot is inspired not by an empathetic human encounter but a nonempathetic form of encounter characterized by the buying and selling of sex." Imagine, she noted, a sex robot that can be programmed, or taught, to deny consent to sex and to resist, only to eventually capitulate. Would it not legitimize and encourage rape culture? And yet how easily can we imagine that many would use a trainable robot to realize such a sexual fantasy? As for Levy's argument that having robot prostitutes would still be an improvement on having human ones, she does not buy that either: "The proposal is that males can be gratified exclusively in the way they desire without any concern for reciprocity and a mutual empathetic relationship. This logic only makes sense if someone believes humans are things, and if they think instrumental relationships between persons are positive with no resulting impact on social relations between persons."

We started this chapter with the fictional story of an awkward but nevertheless happy moment that borrows from Levy's optimistic take on sex robots and paves the way to a more comprehensive emotional bond between humans and robots. But concerns such as Richardson's ones have a lot of merit, and if robotics goes on to enabling some kind of super realistic pornography, or to automating prostitution, they will be more widely debated.

BOSTON, MASSACHUSETTS, USA, JUNE 2, 2072

Their first anniversary was only two weeks away, and the fact that Jack still had no idea of what to give April was making him slightly nervous. She'd mentioned that she had a lot of ideas for his present. *Of course you have*, he'd thought. She could access all of his Web searches to learn about things he was even slightly interested in and could browse and compare thousands of products in a few seconds. Let's face it, he thought: it was much easier for the robotic half of the couple to choose a present for the human half than the other way around.

This morning, though, there was no time to think about her present. They had another present to deliver, one they had chosen together a few days earlier. Almost one year after their own marriage, they were heading to Boston City Hall again, to the very same room that the mayor had chosen for them and that today was hosting the second human/robot marriage in the city. They had met the couple through a common acquaintance—Helen Myzaki herself, the engineer who had made both marriages possible. They had instantly liked each other and began going out for dinner or drinks together. At first, when going out with Eric and Jennifer, Jack and April were mostly enjoying the company of a couple—possibly the only couple—who would not judge them, who would be comfortable in their presence, who would understand. Their first year of marriage had been great: sure, there had been a few bumps in the road and a couple of heated discussions, but in the end they were welcome, as they only went to show that theirs was a real marriage. The really difficult part, besides the constant attention from the media, had been the uneasiness they felt in relatives and friends whenever they got together. Eric and Jennifer's company had been a relief. Soon they discovered they really had a lot in common—beside the obvious—and ended up being true friends. When they finally received the invitation to Eric and Jennifer's wedding, Jack and April were beaming with joy—even more so now that the day had arrived.

Hand in hand, Jack and April arrived at city hall and walked past the usual crowd of journalists, photographers, and curious onlookers. Though not "the" first marriage of this kind, this was still a first, and it was attracting as much media attention as Jack's and April's wedding one year before. They climbed the stairs, entered the room they remembered so well, and smiled as the city officer announced the names of the soon-to-be-wed: Jennifer Masterson, thirty-eight years old, from Hartford, Connecticut, and Eric (from now on also Masterson), her robotic spouse.

6 A DAY IN THE FACTORY OF THE FUTURE

HO CHI MINH CITY, VIETNAM, 2049

It was the first light of dawn when she got out of her car in front of the gray and orange facade of an industrial building in Saigon Tech Park. Tuyen had good reason to be so early that morning. Just like every other day, she liked to avoid the traffic jams that still plagued the route from the city center to the industrial district where her company was headquartered. Ho Chi Minh City had changed enormously during the past couple of decades, claiming its place as one of the most dynamic and fastest-growing high-tech hubs in Asia. But one thing had not changed much: its inhabitants' habit of packing streets with scooters and minivans, making life impossible for pedestrians and for self-driving cars like her own that at rush hour had to crawl along the city's main arteries, their autopilots working at full capacity to avoid running over anyone.

She also had another reason, more unusual, to be early. That was going to be a big day, for her company and for her. That day, inside the gray and orange building, her company would start churning out the product that would take the consumer tech world by storm, setting the company on a path to become a world leader in its field. That was the bet at least—and as the project manager in charge of that product, her career depended on that bet.

She smiled at the face recognition camera to open the door of the manufacturing department, hung her coat, and grabbed a cup of tea from the machine near the entrance, then walked into the main production hall. It was quiet now, empty of humans, with only some beeps emitted by the drones hanging from the ceiling. In less than an hour, the place would be buzzing with activity.

The production pipeline had been tested extensively during the pilot phase, but this was the first day of actual production and shipping, and a lot could go wrong. So, yes, she was nervous. She enjoyed these few moments of calm and silence and waited for her colleagues—human and mechanical—to show up.

Two hours later, that quiet already seemed a distant memory. The production hall was crowded with women, men, and robots. Humans gathered around production islands scattered along the hall, working alone or in small groups, each one focusing on a critical component that would end up in the final product. Robots were in constant movement around the hall. At all times, drones detached from the ceiling in groups of three, four, sometimes six, and flew toward the warehouse. There they went from shelf to shelf, using their onboard cameras to check the inventory, rearranging the distribution of components between shelves, sometimes grabbing small parts and lifting them toward a work cell. Meanwhile, wheeled robots took care of the heavier boxes and equipment, lifting and transporting them from the warehouse shelves across the production hall. There was something mesmerizing about how all these robots kept flying and wheeling around, constantly crossing paths with each other and with humans without ever bumping into each other or having to slow down too much.

But Tuyen had no time to stand in awe of the warehouse robots. She had to follow the first production batch as it progressed across work cells, making sure that the process flowed as planned. She began from the production island at the center of the hall, where her colleague Vinh was putting together the structural core of the product with the help of a coworker.

"How are the two of you going along?," Tuyen asked with a smile.

"Great," Vinh replied cheerfully. An experienced and skilled technician, he had witnessed several launches of new products and would not let the pressure get the better of his good mood. "He still has a lot to learn, but does not lack enthusiasm and never gets tired," Vinh said of his coworker.

Vinh's task was to supervise manufacturing while his coworker carefully inserted sensors, electronics, and actuators into a layer of fabric and stitched another layer on it. The work required precise movements and fine manipulation, and although they had already gone through a few training sessions, Vinh still needed to give some instructions, which he did by using only a few words. He preferred to teach by example. He first showed his partner the careful movements needed to place each component at the right place, then asked the partner to repeat them, gently laying his hands on the partner's wrists and fingers to accompany their movements and slow them down or steer them in the right direction. Another "arm" hanging from the ceiling, which resembled an elephant's trunk ending with two soft, finger-like appendages, had been helping them pick parts and keep them in place.

"There you go—that's how you do it," Vinh said with a huge smile as his apprentice put the final stitches to the first piece of sensorized skin. "In a couple of days, he won't need me anymore, and he'll teach the others," he told Tuyen. "I'll just walk around the production island watching three or four of these things doing all the work, checking on them, and wondering why they don't laugh at my jokes."

Tuyen smiled because she'd enjoyed Vinh's much-needed optimism and because his coworker had just finished the production without a glitch. As soon as the piece was completed, the robot (Why was Vinh calling it "he," now that she thought about it?) closed its soft fingers and retreated its articulated arms. The message "Piece n°1 completed!" appeared on its screen-like face next to a green light, the only sign of satisfaction for a feat of fine manipulation that would be unthinkable for robots only a few decades earlier.

BERLIN, GERMANY, AND LAUSANNE, SWITZERLAND, CURRENT DAY

The first robot to enter a factory was an arm connected to a computer. Unimate, created by robotics pioneers George Devol and Joseph Engelberger, first reported to work in a General Motors factory in 1961, tasked with picking die-cast car door handles, dropping them in a cooling liquid, and moving them toward the welding area, where they were fixed to the car. Six decades later, handling operations (defined as "removing, positioning, feeding, transposing and conveying the work piece or the material") remain robots' most common jobs in factories worldwide, accounting for more than 40 percent of industrial robotics deployments.[1]

More recently, robotic warehouse management has been gaining momentum. In its sorting centers, Amazon uses dozens of wheeled robots, coordinated by central software, to grab packages from the warehouse, group them according to their destination, and place them on conveyor belts that will get them to the delivery vans. The robots are made by Amazon Robotics, a subsidiary of the retailing giant that was once called Kiva Systems and was cofounded by Carnegie Mellon professor Raffaello D'Andrea. D'Andrea, now a professor at ETH Zurich, recently started pilot projects with IKEA, another global firm with massive warehouse management needs, where his latest camera-equipped drones are used to monitor inventories in large warehouses. For the moment, their job is not to handle packages but to upload images of how pallets are distributed among the racks and then go back to a central station. Here the data are downloaded and analyzed by algorithms that produce strategies for using warehouse space more efficiently.

At Amazon and elsewhere, the upstream part of the process—picking products from shelves and packaging them ready for shipping—is still done mostly by hand, but Amazon has been funding "robotic picking challenges" in the hope of increasing automation.

In both manufacturing and logistics, robots handle mostly rigid objects that come in regular shapes. Industrial robots have built their success on handling hardware that behaves in predictable ways when picked up and moved between locations. In 2018, researchers from Nanyang Technological University in Singapore showed that two industrial robotic arms equipped with two-fingered grippers, a force sensor on their wrists, and a 3D camera could reliably assemble an Ikea chair using the popular Robot Operating System.[2] The two robots spent approximately twenty minutes mounting the chair: they recognized and located parts randomly scattered on the floor, planned the manipulation sequence, and eventually assembled the chair. The most difficult part was inserting the wood pins that hold the chair parts together because of the mismatch between the planned and actual movements of the arms. The robots circumvented the problem by gently sliding the pin on the surface until they felt the hole with their force sensors.

Despite this remarkable feat, robots still play a limited role in all industrial processes that have to do with handling soft, fragile, or diverse materials, such as food, fabrics, waste, and other deformable materials. As our EPFL colleague Aude Billard notes, "We are still struggling to have robots replace or complement humans in all places where working with deformable objects is required. We do not understand how to treat the problem mathematically. Also, we have problems with all tasks that require some reasoning, that are not fully repetitive. You have many of those tasks in assembling, or in high-quality production, including assembling drones or other robots themselves." Humans, for example, are very good at precisely inserting objects in the right place, and in particular at inserting deformable objects, which requires a level of fine force control that is currently beyond reach for robots.

Robots with such capabilities would be game changers in many fields, from manufacturing to services. Many industrial processes could make good use of robots capable of grasping and manipulating objects of different sizes, effortlessly switching between small and large ones, autonomously deciding whether an object is best grasped from above with two fingers or wrapped around with the whole hand.

No wonder soft grasping, soft handling, and soft manipulation are hot areas of research. Solving those problems would bring about a second wave of factory automation, after the first one that started with Unimate, and expand industrial robotics beyond the automotive and electronics industry that remain by far the biggest buyers of industrial robots.

German researcher Oliver Brock has been working for years on soft robots for manipulation, trying to understand how to design and control robotic hands and fingers that can approach the level of dexterity of human hands. Brock, who currently heads a Berlin-based institute, Science of Intelligence, believes manipulation is an ideal entry point to understanding it. "If we look at all theories that people have come up with when analyzing the evolution of intelligence, it is entirely plausible to say that human intelligence is about moving in the world," he says. "Everything we do is effectively through motion. At first, we were just floating in water toward nutrients; then a lot of things happened in between and we ended up using tools. But I think that everything is about interacting with the world, and manipulation is the first frontier in interacting with the world. Our brain evolved by taking solutions it had developed to solve a problem, and reusing them to solve more complex problems. Solutions found to satisfy manipulation needs were later reused to solve more complex things."

Several researchers believe that the future of manipulation is soft, and that soft robotic manipulators can combine the best of two worlds. Like classical robotic hands, they can have many degrees of freedom—even more, in fact—that allow a vast repertoire of movements. (Reminder: a degree of freedom for roboticists is one of the ways in which the robot can move its parts. The number of degrees of freedom, especially when referring to arms and hands, can be best understood as the number of joints.) But unlike classical robots, soft manipulators can passively comply and adapt to the properties of soft materials instead of actively computing the precise motion of all those degrees of freedom. They typically do so by inflating, stiffening, or pulling tendons of soft, finger-like appendages that react by wrapping around objects of diverse shapes and materials and gently comply if the objects deform. Soft manipulators can also use other tricks to improve their grasp. Some have small octopus-like suckers activated by pneumatic valves, others have passive gecko-like skins that stick to objects, and yet others use programmable electroadhesive forces, akin to the static electricity that attracts and retains light materials on surfaces.[3] These robotic grippers have many fewer actuators than degrees of freedom, and for this reason, taking a page from the past where each degree of freedom was precisely controlled and actuated, roboticists call them *underactuated*.

A few years ago, Brock and his team revealed a soft, dexterous hand based on pneumatic actuators that he's been refining ever since.[4] Its fingers are made out of soft silicone rubber, except for the bottom side, which is made out of a nonstretchable fiber. Each finger contains an inflatable chamber that changes in size when filled with air, and since the bottom side cannot stretch, the finger bends. Brock's human-like hand is composed of five bending fingers (four long ones plus a shorter thumb) and an inflatable palm that causes thumb opposition, one of human manipulation's strongest assets. To prove his hand's dexterity, Brock had it undergo the Kapandji test, commonly administered to human patients during rehabilitation. The test consists of touching a series of eight points on the other fingers with the tip of the thumb. A human who can touch all of these points is considered to have full finger functionality. Brock's robotic hand, code-named RBO, passed this sort of Turing test for dexterity by touching seven out of eight points. It also managed to execute thirty-one different grasp types on seventeen different objects.

The RBO Hand 2 is not the only soft dexterous hand around. Many others have been created by researchers over the past few years, and many have adopted different principles. But it stands out because it can be easily manufactured in diverse shapes that are suited for specific situations. The manufacturing pipeline, largely based on 3D printing, allows the fabrication of hands of various sizes and grasp strength. Using the desired grasp types and strengths as an input, Brock's team's method selects the appropriate shape and thickness of the individual parts before manual assembly.

Brock notes that we still lack a systematic approach for designing and manufacturing soft robots. If making soft robots is akin to "programming matter," as he likes to say, we have nothing resembling the well-tested, highly formalized methods that allow us to do the same with information—from microchip design to programming languages.

"I don't think it is particularly difficult actually; it's just that we have not been doing it for very long," he says. "It is kind of unfair to compare soft robotics with computer science, for example. That started thousands of years ago, when we began using mathematics to describe physics. The algorithms we're using now may be recent, but they build on a very long tradition." Brock thinks that making progress in this field requires a true zeitgeist shift that is only just beginning: learning to think of matter as something active that can solve problems without requiring computation and control. This paradigm shift is what roboticists call "morphological computation."[5] While classical robotics encodes every single movement of a robot, so that no movement can happen unless it is controlled by a calculation executed by a computer, soft and bioinspired robotics adopt materials and designs that do the work without requiring precise control of all degrees of freedom, effectively reducing the computational load of the computer.

However, manipulation without sensation is not effective. Unless a robotic manipulator is programmed to execute predefined actions at predefined times, which can work only with rigid objects of predefined shapes positioned at predefined locations, some level of sensory information is required for locating, grasping, and securely moving an object. Today, the most sophisticated robotic manipulators on factory floors use video feeds from cameras positioned on the ceiling or on their arms to understand and guide their actions. But visual feedback cannot provide the richness and granularity of touch sensation. In chapter 5, we pointed out how helpless humans are at simple manipulation tasks when haptic feedback is prevented and they must rely only on vision. Human hands are covered with thousands of receptors that give information on static properties, such as texture, pressure, shape, temperature, humidity, and dynamic properties, such as vibration, bending, friction, and deformation. As soft robotic hands become more dexterous, researchers are developing a large range of sensors that can be integrated in soft materials. These sensors leverage resistive and capacitive electrical responses of deforming materials, such as liquid metals and carbon grease, as well as mechanical, magnetic, thermal, and optical signals, to mention a few. Although researchers are still far from approximating the fine sensory capabilities of biological skins, they are making progress in integrating soft sensing technologies that render diverse tactile information[6] and in neuro-inspired wiring methods to connect thousands of such sensors to a single chip.[7]

Designing and building more dexterous and sensitive hardware is only one of the problems that need to be solved in order to crack complex manipulation of objects that today require human workers. Maybe even more challenging is the problem of how to program a robot to manipulate deformable objects. The precise movements required to grasp, hold, fold, and position a piece of fabric or soft rubber band cannot possibly be anticipated and programmed step by step by classical control theory alone. After all, we humans aren't really aware ourselves of how and why we choose to pick up a piece of cloth with two fingers or with a full hand, or of how much force we use in grasping an egg without breaking it.

One option is to use machine learning, a field that has had huge successes in recent years in a wide variety of fields, such as image recognition, automatic translation, and prediction of protein structures. That's how Google Research tackled the problem of picking up objects of different size, mass, and appearance with a two-fingered gripper attached to a robotic arm.[8] The robot brain consisted of a neural network receiving information about the joint angles of the robot and images from a monocular camera mounted over the robot's shoulder that pointed at a bin where several objects were randomly scattered. The neural network was trained to predict the outcome of a grasping action after learning from hundreds of thousands attempts. While other researchers used similar learning approaches in simulation that relied on models of the robot's state-action space, which may not transfer well to real robots or to robots with different geometries, the Google team trained the neural controller entirely on data generated by physical robots; furthermore, they did not build in any prior knowledge about the robot's geometry and kinematics. The team put fourteen robots to work and collected data on 800,000 grasps (equivalent of 3,000 hours of operation of a single robot) that allowed the neural network to accurately predict the best grasping actions for a large variety of objects. Furthermore, the variability induced by the wearing out of the two-fingered grippers, by slightly different camera perspectives and other small hardware differences between robots, allowed the neural network to see diverse situations and generate more robust grasping predictions that were also applicable to other robotic arms, albeit with the same morphology and sensory configuration. Two-fingered grasping is an important developmental milestone for human babies, who learn to pick up objects with thumb and index after approximately seven months of simpler grasping actions. Interestingly, the Google team reported that during the learning process, the robot progressed through the pregrasping behaviors observed in babies, such as focusing attention on the object selected for grasping.

However, our EPFL colleague Aude Billard is a strong believer in the idea that it will never be possible to use machine learning alone. "We will always need to combine it with classical control, for various reasons. One is that you cannot easily transfer what is learned with one robotic platform to a different platform or to a different problem; you basically have to relearn things every time. As much as we love standardization, we will never have just one robotic hand that everyone uses. It is like the dream of all having the same smartphone or the same charger. It will never happen."

More fundamentally, Billard says she would be the "least happy" to see great successes in robotics deriving from machine learning alone. "Science is about understanding, not just having results. What is often happening with machine learning is that people are describing results without being able to explain how they were obtained."

Still, she thinks machine learning could help work out some of the equation variables that engineers would never be able to work out by themselves. In fact, Billard is a pioneer of a machine learning method known as "learning by demonstration." This is different from both classical imitation learning, where the neural network receives the correct answer for many examples, and gradually learns a general strategy, and reinforcement learning, where the neural network tries many possible strategies and receives mathematical "rewards and punishments" depending on the quality of the action outcome.

Learning a new motor skill—be it playing a chord on the guitar, developing a tennis backhand, or fitting a luxury watch's internal mechanism inside its case—is something that can rarely be done by just reading an instruction manual. The best thing is to watch someone who is very good at it and try to copy their movements. Why not try the same with robots then?

Indeed, it can and has been done. A combination of learning-by-demonstration and classical control allowed Billard's group to teach an industrial robotic arm equipped with a four-fingered hand and a set of cameras monitoring the scene from different parts of the room to catch various objects thrown at it: tennis rackets, bottles, and balls.[9] "Combining machine learning and control was the only way to do it," Billard explains. "When we catch a bottle in flight, there is simply no mathematical model that describes how a bottle filled with some amount of liquid flies. The liquid is moving as the bottle rotates; it is simply too complicated to model. But we knew that acceleration, position, and velocity were in the picture—Newton, to put it shortly—and we estimated the other parameters with machine learning." In practical terms, Billard and her team began throwing bottles at people, while filming the scene with cameras. Then they had a machine learning algorithm digest the footage, armed with a generic physical model of each object at rest, in order to learn to anticipate the likely trajectory of each object and to predict how and where a person would grasp it. They then transferred that prediction capability to a simulated robot first, where they conducted additional experiments to expand the training data set, and finally to a physical robot.

Learning by demonstration is rapidly becoming mainstream in industrial robotics. "Almost all robotic arms you see nowadays are backdrivable: you can move them with your own hands and show them the right movement," says Billard, whose group was among the first to introduce this kind of kinesthetic demonstration.[10] The problem, she notes, is that many existing platforms are too imprecise when they reproduce the movement—at least compared to what is needed in industrial production. "What is really lacking is precise application of force. That is very difficult to transfer, and yet manipulation is very much about controlling force," she notes. "We lack good interfaces to transfer forces. We can say that programming by demonstration is a solved problem when it comes to learning trajectories, which is fine but not enough, because industry also needs force control, which is not solved at all."

Still, Billard believes learning by demonstration will get better and will play an important role in factories of the future, especially for reprogramming robots to produce a different object. "Nowadays you need to call in a programmer who comes and reprograms everything, and that costs a lot," she says. "But especially with pick-and-place tasks, you would just want to show the robot that instead of picking an object here, it has to pick it there. And if the new object is smaller than the old one, you just want to show the robot how to adapt its grasp. That would be gold for industry because also untrained workers could program robots."

It will still take much time to crack problems such as soft manipulation, learning, and collaboration. But once it happens, it will force industries to rethink the rigid division of labor that—even in the most automated sectors—separates what robots do from what humans do.

HO CHI MINH CITY, VIETNAM, 2049

After leaving Vinh's zone, Tuyen took a walk through the production hall and spent a few seconds watching the whole scene from the side. She wanted to get the ensemble view, to check if the process was running smoothly or if there were sections where things were too slow or too fast, if bottlenecks could be eliminated. But no, all looked fine: everyone knew what to do and did it well, and all production areas were getting what they needed when they needed it. She took some time to appreciate the racks of long and sinuous cylindrical robots, resembling elephant trunks, that were hanging from the ceiling across the hall. All the time workers were grasping and pulling them toward their work zone, providing instruction with a few touches on their tips and watching them as they curled around heavy cranes to load them on wheeled robots; or as they used their tip to delicately pinch patches of fabric and place them on the production table; or as they blew air out of their tubular structure to clean finished parts before sending them to the next production step. As soon as they were done, those elephant robots curled back into position, a few feet above the ground, and waited for the next worker to ask their help with a gentle knock on their thick, flexible skin.

GENOA, ITALY, CURRENT DAY

Who says manipulation requires hands and fingers at all? As impressive as human handling capabilities are and as crucial they have been for the evolution of our species, other animals can also display a remarkable ability to pick, grasp, and carry things based on completely different principles. And there are cases where their strategy could prove more efficient than ours, so why not add more manipulation methods to our robotic toolbox? At the Italian Institute of Technology, Lucia Beccai is particularly fascinated by the way elephants use their trunk to interact with the world and teamed up with biologists to make robotic versions of the proboscis.

"I think robotic hands are just too complicated," Beccai says. "Prostheses are one thing, but if we need something to help people in factories or to help patients sit on a wheelchair and move to the bed, it's very difficult to give robotic hands the deformation capability, the delicateness, the adaptability you want to have in those situations."

The elephant's trunk turns out to have all those properties, with a deceivingly simple structure. Elephants can curl their proboscis around heavy objects, such as wood logs, and lift them. One minute later, they can use the finger-like extension on the tip to pick a leaf or a flower and bring it to their mouth. The proboscis—physiologically speaking, a fusion of the nose and the upper lip—can also be used as a sucker, typically to draw water but often also as a help for grasping and holding objects on the tip, or for taking up a large number of small objects. And of course, it can work in the other direction and spray water all around. "What is most interesting about it is that it has no rigid support," Beccai explains. "It is a bundle of muscles inside a very thick skin, and that allows the animal to combine movement adaptability with strength."

That combination is indeed quite rare. Many of the same things that soft roboticists find attractive, she says, are exactly the things that make classical roboticists skeptical. "How can I use elastomeric, soft materials and obtain the same strength that I would have with a rigid arm?," people legitimately ask. "That is still largely unknown in soft robotics," she says.

With funding from the European Union and partners from Italy, Switzerland, Israel, and the United Kingdom, Beccai's team is unraveling the mechanics of the elephant trunk and decoding how the animal makes decisions to use the trunk for different purposes. "We want our robot to be strong, yet delicate and precise. A robot with rigid links and joints can be perfect in terms of precision and motion predictability, as automation robots today are, in particular for grasping and releasing objects. But they are not the best solution for manipulating objects of different size, weight, and consistency where different amounts of force distribution are paramount." This means exploring an entirely new paradigm of robotic manipulation. "Instead of a gripper attached to an arm, which uses fingers to grasp and the arm to apply strength, we want a continuum structure where there is no clear distinction between parts that are responsible for fine manipulation, strength, grasp, and release."

Beccai's group is still working out the best technologies that will be used as the robotic trunk, but a few things are clear. They will build artificial muscles inspired by the principle of the muscular hydrostat, a structure shared by the actual elephant's proboscis, the tongue, and the octopus tentacle— another champion of manipulation that combines strength with delicate handling and was extensively studied by Cecilia Laschi from Scuola Superiore Sant'Anna in Italy in one of the first research projects in bioinspired soft robotics.[11] Muscular hydrostats have no skeleton, and the muscle tissue itself both creates movement and provides support for it. The structure is composed of several longitudinal and circular muscle fibers that encapsulate an incompressible liquid, such as water. When the muscles are relaxed, the structure is soft and compliant. Contraction and shortening of muscles in one direction causes a bending in that direction. When all muscle fibers contract, the structure becomes stiff.

A bundle of artificial muscles, made out of soft materials and inspired by this principle, will be embedded within a thick and resistant skin. "There are a lot of beautiful stretchable robotic skins around, but they are very delicate," Beccai explains. "If I want future robots to achieve complex interactions with the physical world, I need something tougher, that can resist high temperature, that can withstand liquid, dust."

The biggest challenge will be to add sensors that will work inside this thick skin. "The elephant's skin is both very thick and very sensitive—in particular the internal part of the trunk tip, which has not been investigated very much even in biology." Beccai's group will integrate sensors for tactile stimuli and proprioceptive stimuli to allow sensing the trunk's position in space and its own deformation, as well as the texture, shape, and features of objects. One of the technical issues will be the wiring: the sensors cannot be at the surface and will have to work under a thick layer while at the same time encoding stimuli about what gets in contact with the surface. The end goal is to build a system that would rely on touch and proprioception to control reaching, grasping, and handling instead of vision. "Reading about elephants, I realized that they prioritize smell and touch over vision," Beccai explains.

Another challenge, as with all other robotic systems, is how to endow them with intelligent control abilities. "We study the behavior of elephants at an African research site, and watch them in their natural environment as they interact with objects," says Beccai. "We use motion capture systems to decode their grasping strategies. For example, we want to understand how elephants decide to curl their trunk around an object or grasp it with the tip. What are the criteria, in terms of object's size and shape, used by the elephant to decide how to manipulate the object?" In the end, they will look at footage and motion capture data and try to extract stereotypical behaviors, standard reaching and grasping strategies, that the animals apply consistently for similar objects and then encode them as control strategies for the robot.

Beccai sees many potential applications for a robotic system that can truly reproduce the elephant's proboscis. "In manufacturing, for example, in the automotive industry, I can imagine racks of proboscis-like tools that hang from the ceiling. The worker would grab and extend them and use them to lift heavy parts or to help manipulate small objects. And let's not forget the internal cavity, which can be used to provide air, water, or other liquids required by a manufacturing process."

HO CHI MINH CITY, VIETNAM, 2049

Phuong, the young engineer in charge of the next production stage, was a very different companion than Vinh was. She did not crack any jokes, and she answered Tuyen's questions with few words, hardly taking her eyes off the robots that were putting the finishing touches on the almost completed products, which would be ready to ship in a couple of hours.

But Tuyen did not like her less for that. In fact, she saw something of herself in her—the same drive and attention to detail—and she thought she was going to become a really good manager at some point. She had full trust in her, which is why she had put her in charge of this critical step.

Under Phuong's watch, the finished products were taking shape. On one side, a group of dexterous robots were cutting and modeling foam to create cylinders, ovals, spheres, and other irregular shapes that were going to constitute the internal structure. Other robots carefully picked the high-tech fabric assembled in the previous step by Vinh's coworkers, cut them into various sizes and shapes and wrapped the pieces around the foam structures.

In watching this operation, Tuyen was reminded of a trick she loved when she was a child—those inflatable balloons squeezed, stretched, and tied together to create models of animals. Here, too, skins and foam were being assembled into shapes resembling a cat, a puppy dog, a snake, a bird. But the real magic trick was not in the shape. It was in what happened after they were wrapped in the skin. As Phuong clicked a button on her tablet, each one of them began moving, walking, jumping, crawling—and sometimes changing shape in the process. In a few minutes, those humble, dumb, and uninteresting parts had been turned into robotic pets thanks to their robotic skin.

In the end, there they were: the new products that were about to start shipping from the Saigon factory: a new type of soft robotic companions built by robots and humans. They were going to be "the" present of this year's holiday season—at least judging from the preorders. They had something for the whole family: for children, they would be halfway between toys and pets that would play with them while also having enough intelligence to keep them out of trouble. For grown-ups, they would be pets doubling as personal assistants that could fetch and deliver the newspaper to the couch while keeping an eye on the house—helped by a wireless connection to an alarm system. And for the elderly, they were going to be distractions, health sentinels, and a way to connect with grandchildren.

The ease with which the production process churned out such a variety of shapes—dog, cat, bear, koala, parrot, baby dinosaur—was due to the robotic skins that wrapped the core and morphed into the desired shape and functionality. Instead of changing the manufacturing procedure for every toy variation, Tuyen could rely on the programmable robotic skin to transform the core into the desired toy. It was almost as easy as coming up with a new balloon trick.

And toys were only the beginning in Tuyen's plans. The same skins could be used to create a whole new world of animated objects—tools, lamps, pillows, bags. At some point she hoped her company would start selling them

as stand-alone products, as the most tech-savvy users would learn to program and apply them to whatever soft object they had in their house—realizing a vision first set out by an American researcher in the late 2010s.

NEW HAVEN, CONNECTICUT, USA, CURRENT DAY

As science journalists know well, sometimes it can be difficult to get scientists to describe their work briefly and in lay terms. But that's not the case of Rebecca Kramer-Bottiglio, a professor of mechanical engineering and materials science at Yale University. When it comes to summarizing the main point of her *Science Robotics* article in 2018, "We can take everyday objects and turn them into robots," she says.[12]

Kramer-Bottiglio's goal is to make robots more versatile. Robots can have many different applications, but they typically do only the one thing for which they were designed. Reconfiguring a robot to do something else is very difficult, yet it could be critical in many cases. "The project started as a collaboration with NASA," she recalls. "We were thinking about how robots could adapt to extraterrestrial environments where we don't know what to expect, and wondering: How can we make robots resilient enough to survive in such an environment?"

The cost of bringing to space multiple robots that can perform multiple functions is very high, so the researchers wanted to create a technology that could apply to whatever object is available. The proposed solution was a flexible skin, with actuators and sensors sandwiched between two sheets of fabric. In order to produce different movements, they tried two different types of actuators: pneumatic ones, which produce movement by injecting air into inflatable chambers (similar to the ones Brock used for his soft hands), and other actuators based on shape-memory alloys, which are materials that can deform in predictable ways when an electric current is applied. In both cases, the actuators were paired to strain sensors that could measure the deformation produced in the skin, enabling a closed-loop control system where feedback from the sensor can be used to adjust the motion of the actuators. The skins were then wrapped around soft foam cylinders, leading to a crawling inchworm, a pair of legs, three gripping fingers. Because the sensors measure only the deformation of the skin, the control system works independent of the material around which it is wrapped.

Kramer-Bottiglio's group then stepped up their game, moving away from foam-based deformable objects to moldable materials.[13] This time the skins did not just bend or squeeze a cylinder that would always remain a cylinder; it was wrapped around a ball, made out of a modeling compound for children, and turned it into a cylinder and then into a bell-like shape. "The skin acts like a surface sculptor," she explains, "applying surface strains and pressures to mold the volume into a completely different shape." The idea, she says, is to create shape-changing robots "that can adapt not only their behavior but their whole shape to changing tasks and environments." The robot could start out as a ball in order to roll fast toward its destination, then turn into a cylinder and adopt a worm-like locomotion to pass through a hole, and then change shape again as it comes to a rock that it can't roll over.

Future challenges lie in improving skin dexterity and cognition, Kramer-Bottiglio explains: "Right now we are limited to radially symmetric shapes, such as spheres, cylinders, bells, because in order to apply sufficient force, we need the robotic skin wrapped all the way around the shape." In the future, they hope to replace pneumatic actuators with new actuation technologies that can deliver more force at smaller size and also operate with soft cores of different shapes. As for cognition, "We have this vision of robotic skin being very resilient, but soft sensors and electronic components tend to be very delicate, so there is a mismatch." The group is developing soft, stretchable sensors and electronic circuits embedded in the skins.

The applications of robotic skins would go far beyond space. Kramer-Bottiglio notes, "There are a lot of unpredictable environments on Earth as well—search and rescue is a typical example. A robot that can change its shape and locomotion mode on the fly could crawl through rubble and look for survivors." The concept may also be applicable to wearable robots. For example, her group showed a shirt-like device to correct bad posture by arranging the skin actuators in a triangular fashion, a configuration that allowed them to adapt to diverse body curvatures.[14] Kramer-Bottiglio says that she has been approached by toy companies: "They see this as a potential

Lego-type kit. People could buy a kit of robotic skins and apply them to whatever soft objects they have in their household to design robots on the fly."

It is no coincidence that Kramer-Bottiglio and her team chose to wrap their skins around the four legs of a plush horse for a demo video that accompanied one of their publications.[15] In future manufacturing, such robotic skins could be part of the final product—as we imagined here—or of the manufacturing process itself. For example, they could help quickly reconfigure and reshape a gripping device: multiple soft fingers or trunk-like elements could be assembled in various configurations and wrapped in robotic skins with different orientation in order to shift from a three-fingered hand to a five-fingered one, or to add an extra thumb, or to combine hand-like and proboscis-like manipulation.

As for final customers, the possibilities would be much wider than toys and robotic pets. One can imagine people buying (or 3D printing themselves) soft cores of various shapes, as well as loads of patches of all sorts of robotic skins—triangular, rectangular, big, small. Downloadable apps would help them in assembling and configuring the robots. The same components could be used one day for a weight-lifting support system to help moving stuff from, say, the basement and be reassembled the next day into an exercise machine, or a massaging chair. Kramer-Bottiglio's idea could be one of the most promising routes to the vision of "a robot in every home," except that instead of really buying *one* robot, people would be turning what they already have in their homes into robots.

HO CHI MINH CITY, VIETNAM, 2049

It had been a long day. Most production islands were already quiet and empty when Tuyen took the last batch of toys—a horse, a parrot, and two octopuses—in her own hands, placed them in a cardboard box, and sealed it with duct tape. A robot had been doing this trivial operation for the whole day, but she had pushed it away to do it herself for this last one.

Apart from a few glitches, the day had been a success. The production targeted for the day had been surpassed. All the products had undergone the standard functionality tests before exiting the premises, and less than 1 percent had to be discarded. The following day would be just as challenging because the production target was going to be higher. But for the moment, Tuyen could walk to her car, head home, and relax. Unlike at the start of the day, she would not be able to avoid Ho Chi Minh City's traffic, but at least she would have company: the little bear that had been the first finished piece to come out from Phuong's table and that she'd decided to keep for herself, was now stretching its legs on the passenger seat beside her.

7 THE FIRST NOBEL FOR ROBOTICS

STOCKHOLM, SWEDEN, DECEMBER 8, 2052

This is an excerpt from the Nobel lecture delivered in the Aula Magna of Stockholm University by Professor Valérie Paquet, one of the winners of this year's Nobel prize for medicine for "the development of bionic limbs and exoskeletons that allow paralyzed patients to walk again."

Your Majesty, fellow awardees, distinguished colleagues, and all of you in the audience, here and online: good evening. It is an honor to be here and to share the stage with the other Nobel prize winners. I am so in awe of the work of all of you: the amazing computational chemistry that led to synthesizing new catalysts that are now in fuel cells everywhere and that have revolutionized the energy market, fully deserving the chemistry prize; the study of the role of cryptocurrencies in financial crisis, acknowledged by the economics prize; and of course the physics prize for the discovery of the true nature of dark energy. Let me just say that I have been an amateur cosmologist all my life, and it is just great to be on the same stage with the people who finally solved the dark energy puzzle. It took a while, and in the end it was something very different from what people thought when I was a kid (*laughs*), but thanks for that.

I was told that this is the first Nobel prize ever awarded to robotics and the first prize that recognizes research work largely done by a private company—in this case, the spin-off that I started upon finishing my PhD. Indeed, up to a few decades ago, robotics was hardly considered a science in its own right. It has existed as a discipline since the 1960s at least, but until the early twenty-first century, it was considered just another branch of engineering or computer science. What robot-icists did, in the eyes of many people at the time, was to apply well-understood physical knowledge and use it to build reliable systems that behaved in a pre-dictable way. The most challenging part of their job, in this view, was integrating components into a functional system, without major "discoveries" involved. Rather than answering fundamental questions about the world or discovering previously unknown facets of reality, as my cosmologist friends here did, roboticists were often seen as tinkerers who put together pieces of well-understood technology, be it computers, electric motors, cameras, or software.

I do not have time to delve too deeply into the history of robotics here, but this started to change between the end of the twentieth century and the beginning of the twenty-first, when roboticists began to look beyond industrial applications and started to conceive life-like, or bioinspired, as they liked to say, machines. The moment they stopped settling for robots that could only execute preprogrammed and repetitive behavior and began shooting for machines that could sense the outside world and learn from experience and imitation as living organisms do—at that moment, they realized they had to tackle fundamental questions about how living organisms work, and in particular how their body and nervous system are intertwined, because biology and neuroscience did not have all answers. Robots became both a motivation and a tool to study living systems, gaining new knowl-edge, and the question of how to transfer this new knowledge into new engineering principles became a science in its own right.

The shift was evident in all branches of robotics, including, of course, the design of robots that would have to work in symbiosis with the human body, at the point of becoming part of it. And this is the story I am about to tell you today. I will speak about the work of pioneers who allowed me to be here today rather than my own work. By the way, it is a pity that the Nobel prize still has not changed the rule by which no more than three people can share the prize because many of them should be here today.

Christians in the room will know that the miracles attributed to Jesus or to the saints often involved making the paralyzed walk again. And indeed, restoring move-ment after spinal cord lesions has long been a dream of regenerative medicine. Each year, according to the World Health Organization, more than half a million people worldwide suffer from a spinal cord injury that severely impairs their ability to stand upright, walk, grasp, conduct most daily activities, and have a sex life—in other words, lead a normal life. Neurodegenerative diseases and cerebrovascular accidents, whose prevalence has skyrocketed in the last thirty years due to the—otherwise very welcome—increase in life expectancy, also often deprive patients of the functionality of their arms, hands, or legs.

Medicine has always resorted to technological devices to help such patients, crutches and wheelchairs being the most obvious examples. In the early 2010s, three scientific trends began to converge that would lead to today's robotic rehabilitation technologies. First, neuroscientific studies suggested that, contrary to the prevailing wisdom from the previous century, it is possible to repair neuronal connections or to get them to heal themselves after a lesion. It is difficult and slow but possible in principle. Furthermore, they discovered how to read out neural signals and stimulate neurons with signals that the peripheral nervous system could understand. Second, computer scientists made significant progress in creating computational models of human bipedal walking, a problem that had been haunting designers of humanoid robots for decades. Third, progress in material science and soft robotics brought a range of actuators and sensors that could be integrated into rehabilitation devices that looked and felt like garments.

It was time for neuroscience, computer science, and robotics to join forces: sensors, electrodes, and brain-computer interfaces could be used to harness whatever signal was left from the nervous system and muscles of a paralyzed patients and, with the help of machine learning, turn them into a language between the brain and a new generation of soft robotics prostheses. And these, in turn, could adapt to the progress of the human and allow patients to undergo rehabilitation protocols in the comfort of their homes while their newly stimulated neural connections started to heal.

The amazing acceleration in wearable technologies that allows me to stand here today and accept this prize happened at the intersection of those technologies and has led to a whole family of soft and light exoskeletons, neural implants, and brain-machine interfaces that today help thousands of patients—from the partially disabled to the fully paralyzed—to recover at least part of what they've lost. In the best cases, they can walk again almost normally. In many others, they are at least able to walk with crutches and move around on special bikes instead of wheelchairs, which would have been their only option up to a few decades ago.

I should take this opportunity to mention a few pioneers who, in the first decades of this century, did ground-breaking work that allows me to be here today.

The first name that comes to my mind is Conor Walsh, a brilliant Irish roboticist who has been for years the John L. Loeb Associate Professor of Engineering and Applied Sciences at the John A. Paulson Harvard School of Engineering and Applied Sciences and a core faculty member at the Wyss Institute for Biologically Inspired Engineering.

Conor pioneered the transition from hard to soft exoskeletons. Exoskeletons, along with bionic arms, hands, and legs, had already been around for some decades at the time. By the mid-2010s, a few dozen medical exoskeletons were on the market for lower or upper limbs, stationary or mobile, targeting human rehabilitation or augmentation—in other words, making daily activities easier for people who had no hope of recovering their functionalities. But most of these devices were heavy, difficult to use without medical supervision, and expensive.

The motors, sensors, and mechanical components of early exoskeletons derived mostly from technologies developed for factory settings, the industrial robots that had made the fortune of the robotics industry since the second half of the twentieth century. When a small group of researchers at Harvard—including Robert Wood, George Whitesides, and Gene Goldfield—started working on soft robotics in the early 2010s, Walsh saw the opportunity to apply this emerging subfield of robotics to the design of exoskeletons—or, better, exosuits, where light fabrics replaced rigid elements of conventional exoskeletons. On the other side of the Atlantic, Walsh was soon joined in this effort by Robert Riener, a German biomechatronic specialist based at the Swiss Federal Institute for Technology in Zurich, Switzerland. The two often collaborated, while each developing their own version of exosuits for rehabilitation and assistance in daily activities. I do not have the time to tell you more about their work, but I encourage you to read their papers.

BOSTON, MASSACHUSETTS, USA, AND ZURICH, SWITZERLAND, CURRENT DAY

The winner of this imaginary Nobel prize of the 2050s may indeed not have time to delve into the details of what is going on right now, in the early

2020s, in the fields of wearable robotics, robotic prostheses, and rehabilitation robots. But we do want to take a closer look at Conor Walsh and Robert Riener, who have been working for some years on light exoskeletons—better: exosuits—designed to help their users stand, walk, climb stairs, and run.

Born and raised in Ireland, Walsh has said in interviews that he got interested in wearable robots after reading a 2007 *Scientific American* article about early exoskeletons developed by DARPA, the research agency of the US Department of Defense.[1] Indeed, the very concept of robotic exoskeletons was pioneered in the military sector, with the idea of using such devices to help soldiers walk long distances with heavy loads. But the same core technology can in principle be applied to the medical sector, for helping people who have partially or totally lost the ability to walk, or in the professional sector, to help workers who have to carry loads or stand for long hours. And Walsh's work does indeed cover all three application fields: military, the workplace, and the clinic.

Realizing that there was not much happening about exoskeletons in Ireland, after graduating from King's College in Dublin, Walsh applied to a bunch of places in the United States and ended up at MIT. There he hooked up with Hugh Herr, an almost heroic figure in this field and for a long time the director of the biomechanics research group at the MIT Media Lab, who became Conor's first mentor. After a hiking accident led to the amputation of both legs below the knee when he was eighteen, Herr gave up his dream of becoming a professional hiker. Though he would soon return to hike, he began an academic career in engineering, specializing in making better prostheses for people who shared his problem. Twice he was on the list of the Top Ten inventions that *Time* magazine used to compile at the time—first, in 2004, with a computer-controlled artificial knee that could sense the joint's position and the load applied to it, and again in 2007, with an ankle-foot prosthesis that for the first time gave amputees what looked like a natural gait. In 2004, Herr also became interested in exoskeletons; he put together a research group and hired Walsh.

Some years later, in an interview with the robotics website RoboHub, Conor would recall the unique interdisciplinary atmosphere of the group, where artists and musicians worked side by side with engineers. It was in this environment that Walsh and others began to think about a new generation of soft, wearable robotic devices that would overcome some of the limitations of existing machines.

As our fictional Nobel prize winner remembers in her speech in 2052, there are quite some exoskeletons around here in the 2020s in various development stages, and a few are available on the commercial market. In 2016, the California-based company Ekso Bionics received clearance from the Food and Drug Administration for a rehabilitation exoskeleton, the Ekso GT, approved for use with stroke and spinal cord injury. A few other companies sell similar products, which can weigh from about 20 kilograms (the weight of the Ekso GT) up to 40 kilograms, and cost several tens of thousand dollars. These weights mean the devices cannot be carried around all day to support daily life activities, and price is a limiting factor too. As *Scientific American* put it in 2017, "The companies that make them typically live from grant to grant . . . hoping to sell rehabilitation exoskeletons in onesies and twosies."[2]

DARPA, the research agency of the US Defense Department, keeps funding efforts in this field. A few devices have been developed by big defense contractors such as Raytheon and Lockheed Martin, but none has yet been employed in the field. Meanwhile, some of the companies producing rehabilitation exoskeletons are adapting their technology to products aimed at the professional market, such as the EksoVest, a passive device designed to provide support to the arm and torso for assembly-line workers who perform overhead tasks such as screwing bolts while standing underneath the vehicle. In 2018, seventy-five such vests were assigned to workers in Ford plants. Another example is a line of soft "powered clothing" commercialized by Seismic, a spin-off from a research program conducted with DARPA funding at SRI International in the United States. Seismic adapted technologies originally designed to give extra resistance and strength to soldiers, into a light suit to be worn over clothes to provide additional strength and support to the lower back. And in San Francisco, the start-up Roam Robotics—founded by Tim Swift, one of the inventors of the first Ekso—develops systems made up of a knee orthosis plus a backpack that help relieve knee pain, reduce fatigue, and even improve skiing.

Walsh is working to push the envelope of these technologies. He aims first for lightness: his devices are between 4 and 5 kilograms. How does such a lightweight device manage to transmit significant force to the user? Walsh's exosuits have a garment part, essentially elastic straps that can be fitted around the patient's waist, thighs, and calves. A set of wires runs through strategic points of the garments, and motors (stored in either a backpack or at the waist) can pull or release the cables to reduce or increase the distance between two attachment points on the suit—not too different from what happens when a muscle contracts, in effect becoming shorter and pulling the bones in the desired direction around the joint.

Spandex, cables, and pulleys may not seem like sophisticated machinery, but exosuits also include sensors—in particular, inertial measurement units (IMU), that detect a body's position in space and force sensors that can measure the force being applied by the trunk and the patient's own muscles. The goal is to detect the person's movement intention and get the cables in the exosuit to move accordingly. Unlike in previous, heavy exoskeletons, the job of keeping the body's stability is mostly left to the patient's bones, as it should be. But the device provides the force needed to move the articulations to the required position at any time.

Walsh began developing the concept around 2014 and perfected it throughout the late 2010s, working both on clinical and nonclinical applications. "Of course there are differences in how you control the device in the two settings," he explains. "For our work on healthy augmentation, it is really important that control works in sync with the user's natural movement. If someone uses an exosuit to go hiking, for example, she wants to be helped but they do not want their gait to be changed. The same applies when you use an exosuit to reduce the risk of injury for people. You can have people who are healthy, but lift boxes all the time and risk hurting their back. Again, we do not want that person to feel different when they are working." Things can also change for people whose natural gait has already been altered by injury: "Patients typically have a specific gait impairment, and you may be encouraging them to try and walk in a slightly different way, for example more symmetrically, so that they can get less tired and exercise more efficiently."

In a series of studies with healthy volunteers, Walsh showed that his exosuits could reduce the metabolic cost of walking. In a 2018 study, for example, he had two volunteers walking the same country course twice, first with the exosuit actuators switched on and then off.[3] He used indirect calorimeters, which are sensors that measure the concentration of oxygen and carbon dioxide in someone's breath, an indirect indication of how much energy that person is spending. The results were encouraging, showing a 7 percent reduction in energy spent for one of the subjects and almost 17 percent for the other.

Why the difference? No two persons walk alike. The same year, his team came up with an algorithm for quickly adapting the exosuit's control system to the specific walking style of each person to ensure that the forces applied by the cables were appropriate to obtain the best energetic efficiency for that user. Personalization is an important feature for turning exosuits into devices that a large number of patients could use. Whereas mass-scale personalization in the garment industry translates into manufacturing clothes in different sizes, future exosuits will also require wearable artificial intelligence that seamlessly translates motion and body signals from the wearer into suitable support forces for different body types, circumstances, and behaviors.

"The idea is that while you are walking with the device, it monitors how you are moving and how your body is behaving," Walsh explains. "It could be monitoring your kinematics, the strain on your muscles, how much energy you consume. Then a control strategy will try to optimize some of the control parameters to help minimize or maximize some of those objectives." Just like automatic translation becomes ever more precise as more people use it, feeding more data and examples to the machine learning algorithms that power it, Walsh expects the control algorithms of exosuits to become more efficient as more and more users wear them. "The system may monitor how people move when wearing the device, build a large data set from many users, and learn how to select what approaches give the best assistance to people depending on their specific impairment," he continues. Machine learning could also play an important role in actually reducing the need for onboard sensors, notes Walsh: "It's very hard to measure everything about a person's function or response to a wearable device with wearable sensors. Often the equipment needed to do this is in a research laboratory environment where you have masks for measuring oxygen consumptions or cameras for motion capture. In the future, we hope to use machine learning approaches where we collect large data sets of how people are actually moving, and we may be able eventually to estimate those things using a minimum number of sensors on the person and combining their readings online with large data sets."

You will have noticed that that we have only mentioned walking so far. It's the most used, but not the only way, for humans to use their legs. We can also run, a fundamentally different style of locomotion; it is not only faster but more energy efficient above a certain speed threshold.

Attempts to build exoskeletons to assist running have often failed, to the point that experiments have shown that more energy is required to run with an actuated exoskeleton than without. The disadvantage of carrying around the weight of the device when running is greater than the help provided by the actuators. In an intriguing study published in *Science* in 2019, Walsh showed how an exosuit can automatically switch between the two very different settings required by walking and running, reducing the metabolic cost for both cases.[4] The suit—this time only assisting at the hip—was equipped with an algorithm that could use data from the sensors to understand whether the user is walking or accelerating toward running and then switch the actuation pattern accordingly. The reduction in metabolic rate could be compared to the effect of taking off 7.4 kilograms at the waist for walking and 5.7 kilograms for running.

More recently, Walsh revealed a lightweight and soft exosuit combined with a few integrated rigid components that can provide help to stroke patients or individuals with cerebral palsy, specifically supporting the function of knee extensors but otherwise totally transparent to the wearer.[5] Six people tested the exosuit walking uphill and downhill on a treadmill, confirming that the device reduces the work that the knee extensors have to do in order to allow walking.

Walsh imagines a future where some wearable devices, in addition to helping patients and workers in specific occupations, become consumer-level products: "I think the same thing that happened in the drone industry will happen in wearable robotics. Drones were premium, high-end expensive systems in the beginning, but then the cost dropped quite rapidly. Wearable robots are more complicated because they interface with the human body, but the core technology is not so different. And for some simple systems, such as devices strapped to ankles or to the back, we are not that far away from those devices being something a person can just buy and use like you would buy any other tool. But that will need a leap in performance, and that will not come from only one technology sector. Sensors, actuators, garment, control will all have to make progress."

Whereas Walsh works on both clinical and nonclinical application of his exosuits, Robert Riener, with whom Walsh has frequently collaborated, is more focused on health and wants to create exosuits that can help disabled—or simply ageing—people in everyday activities. Such a device would be worn during the day, and its actuators need to be geared toward supporting the body's weight against gravity, in addition to helping with moving legs forward. In these situations, upward forces are even more important than forward forces, which are the focus of Walsh's work, and helping the knee becomes more crucial than helping the ankle. Riener's Myosuit, as his exosuit is named, adopts the biological analogy even further, with a three-layer architecture directly inspired by bones, ligaments, and muscles that make up the human walking machinery. The first layer, the "bones," is the actual garment. It resembles a pair of trousers complete with a belt, which provides support to the actuators and sensors and includes stiffer, corset-like sections at the thigh to provide additional stability. The second layer, the "ligaments," is made of rubber bands that connect the belt with the thighs, and the thighs with the shanks. The third layer, the "muscles," is made of cables pulled by electric motors and pulleys. Similar to Walsh's exosuits, a set of sensors continuously measures the angle of the joints, the position in space of the belt, and the tension being applied to the artificial tendons. The control unit uses these measurements to calculate how much actuation is needed at every moment to counteract gravity for a simulated intact leg and move the electric motors accordingly. This way, the suit can help its user sit or get up from a chair and climb stairs, and, crucially, patients can use it to accelerate and improve the quality of rehabilitation. Recently, Riener started collaborating with other Swiss researchers to combine his exosuit with other rehabilitation technologies that target directly the source of the problem for patients who can no longer walk: the neurons in the spinal cord.

STOCKHOLM, SWEDEN, 2052

The work that Walsh and Riener initiated was carried on—by themselves, as well as by other researchers and private companies that soon joined the race—over the following decades, resulting in lighter and more powerful exosuits, thanks to progress in soft materials and miniaturization of components. Similar systems for the upper part of the body were also developed. With more data from users, researchers were able to run machine learning algorithms on their systems, greatly improving the control of exosuits. Now they can be easily personalized to each individual user with just a few hours of calibration. They can pick even the subtlest clues from your stance and your joints to understand in what direction you want to move, how quickly, and how much help you need. People who have to stand, walk, or climb stairs all day for a living—hairdressers, nurses, security guards, movers— often find great help in them.

Initially, the same systems were used for rehabilitation and for assistance in daily life. But the two fields diverged as the market expanded and as scientists gained a better understanding of the different control strategies needed for using an exosuit in a clinical setting or at home. Increased production and competition led to economies of scale and brought down the price of exosuits, to the point that even young, healthy people began to buy and use them.

Yet the ultimate goal of applying robotics to this field was always helping even the most seriously impaired patients—the ones who have not only lost force in their muscles but whose nervous system is no longer able to send out the necessary signals to control movements—both voluntary and involuntary: patients who are not able to lean forward or to bend a knee to signal their intention to walk, sit, and stand; patients with paralysis from the waist down or even from the neck down; patients who lost their limbs in accidents.

The early decades of the twenty-first century saw researchers plant the first seeds of the results that allow me to stand here today, accepting this award. Researchers such as John Donoghue at Brown University did pioneering work on brain-computer interfaces that let paralyzed patients control robotic prostheses.

Others, like José Millan, learned to combine brain-computer interfaces with electrical stimulation in order to promote recovery in patients who would not otherwise be able to control their hand exoskeletons. Silvestro Micera and others wired sensory feedback from bionic hands to the peripheral nervous system, making humans feel again tactile properties of the objects and making the control of the robotic hand more natural. Many of these researchers joined the field of rehabilitation robotics from different fields: they were neuroscientists and clinical neurologists rather than engineers. And they saw robots as a tool to help the reorganization, regeneration, and retraining of the nervous system itself.

At the time, studies in animals were showing that promoting some level of regeneration of damaged spinal neurons was possible. Various strategies were being tried out, ranging from biological—drugs, cell grafts, implants with biomaterials—to technological—deep brain stimulation, direct electrical stimulation of spinal neurons, transmagnetic stimulation, brain-machine interfaces coupled with robotic prosthesis.

One thing was clear enough: even in those pioneering days, there was no magic bullet, and the solution to the problem of neural recovery—if there was a solution at all—was to be found in a carefully balanced mix of different technologies.

Between the 2010s and 2020s, the work of Grégoire Courtine, a French physicist turned neuroscientist, was crucial in showing that this was indeed the way forward for the treatment of spinal cord lesions. Going all the way from animal experiments to clinical studies on humans, Courtine first showed how to combine trains of electrical stimuli delivered to the spinal cord at precise times and locations—rather than continuous and widespread electrical stimulation that had been tried until then—with a robotic system that helped patients fight gravity without interfering with their own leg control. In a memorable study published in *Nature* in 2018, Courtine's group reported how this strategy allowed three paraplegic patients to regain voluntary control of leg muscles that had been paralyzed for many years.[6] After an extensive and painstaking study of how spinal cord neurons are matched to leg muscles and how they transmit signals from the brain to the limbs in healthy subjects, Courtine's team could implant an electrical stimulator next to the spinal cord of the three patients and use it to deliver short electrical bursts that in all effect mimicked the signals that would normally come from the brain to regulate walking. The three patients were thus able to enter a rehabilitation protocol.

With the help of gravity-assist suspensions attached to a robot hanging from the ceiling, they quickly learned how to synchronize their pace with the frequency of the stimulation and walk with the help of crutches. After months of training with the electrical spinal stimulation and the gravity-assistance robot, all three of them had recovered some voluntary control of their legs without further need of electrical stimulation. Two could transit from sitting to standing and walking independently with crutches and could sustain a full extension of their previously paralyzed legs against gravity, and one could even take several steps without any assistive device.

It was a proof of concept, but it was a breakthrough—and a paradigm shift. For neurotechnology in general, it showed that targeted stimulation was the way forward. And for rehabilitation robotics, it showed that it could create the conditions for nerves and muscles to work again rather than replacing them altogether.

GENEVA, SWITZERLAND, CURRENT DAY

It took a while for Grégoire Courtine to decide what to concentrate his hyperactivity on. He had had a passion for science and technology since his early childhood, turning his room into a little physics laboratory and spending hours coding his own computer games on a personal computer. He started playing piano when he was five years old and remains an avid classical music and opera fan. He practiced many sports, including rock climbing. When it came to choosing a subject for his university degree, he chose physics, dreaming that he would become an astrophysicist. Until a chance encounter with a professor convinced him to pursue a master in neuroscience after his physics degree, he did not have any specific interest in studying how the nervous system controls movement—and how to restore that control when it's lost. But in retrospect, it all made sense. All of his other passions proved connected to what would become his career's focus. Piano playing is a good example of what training and repetition can do to the brain circuits that control movements. Rock climbing is a spectacular example of the interaction between the physics of the human body and gravity and of how the neural control of movements works at their intersection. As for physics, far from being a false start, it provided him with the right toolbox for approaching the study of the nervous system. "Physics makes you very comfortable with data and complex processes," Courtine now explains. "It prepares you well virtually for any kind of science, particularly when you want to understand complex systems. At the end of the day, the central nervous system is a complex system like any other."

After completing a master's program in neuroscience, Courtine abandoned plans to study stars and galaxies and went on to get a PhD in experimental medicine from the University of Pavia, Italy, and INSERM Plasticity-Motricity in France. In the early 2000s he was a researcher at the Brain Research Institute of the University of California at Los Angeles. He went to Switzerland in 2008 and since 2012 has been a professor at EPFL, where in collaboration with neurosurgeon Jocelyne Bloch at Lausanne University Hospital, he produced some of the most spectacular results in experimental treatment of spinal cord lesions, including that 2018 study on three paraplegic patients that made headlines around the world.

Although Courtine is part of the rehabilitation robotics community, robots for him are a means to an end. "The key to my work is activity-dependent plasticity. We want to engage the nervous system in order to promote regeneration and maximize recovery," he clarifies. "And the way to do it is to apply what I call an ecological approach to prosthetics. You want to create the most natural conditions possible, and often the rehabilitation environment does not provide that. Robotic systems have enabled me to approach this natural condition in terms of interaction with gravity, of making the patient feel like he is really walking, doing the movements he would do in real life, which is key to promote recovery."

Robots and algorithms, in other words, are tools he uses to help neurons fight back. After a stroke or a spinal cord lesion, the communication lines between motor neurons and muscles are typically muffled, if not totally interrupted. Most patients can still move, thanks to a few neural circuits that have survived the lesion. But like a football team that has to change its plays after a key player is expelled, those circuits have to reorganize. Our nervous system spends the first years of our life fine-tuning the complex interplay of signals that have to travel back and forth along thousands of neuromuscular connections in order to continuously balance the dynamics of our body with gravity's pull—in a word, walk. If it has to accomplish the same thing with only a subset of those connections, it needs to work out a new arrangement. Patients with partial lesions literally have to learn to walk again. As of 2022 they normally do it walking on a treadmill while supporting themselves with their hands on a bar, using crutches, or having rigid supports applied to their legs to help them stand upright. But in all these cases, the interaction between gravity and the patient's body is very different from the natural one—the one that the nervous system had learned to master before the injury.

Robotics can reestablish that natural interaction with gravity, making rehabilitation exercises easier, more effective, and more similar to how walking works in real life. When he set out to include robots in his work, though, Courtine was not satisfied with the typical exoskeletons used for assisting paraplegic patients. "They work by denying the very mechanisms that the nervous system uses to control movements," he laments. "The spinal cord is a very smart information processing interface. Not as smart as the brain, but still smart; it makes a lot of decisions. And it does it based on sensory information." For the spinal cord, Courtine notes, the foot is like a retina that is looking at the world. Proprioception informs the system about the position of your limbs in space. The spinal cord integrates this information and knows where we are, what are the constraints, where we should go next. If a robotic system does not respect this sensory information, it provides the wrong input to the spinal cord, which will interpret it as if there was a problem. "When the robot takes the foot and pushes it up during the swing phase (when the foot swings toward the next foothold without touching the ground), it is like telling the spinal cord, 'We are in the stance phase' (when the foot is on the ground). The information you give to the nervous system is the contrary of what's happening. It makes no sense." And it makes it very hard for the nervous system to learn, because in order to learn, it has to understand what's happening.

So, for his experiments, Courtine went for a very different robotic design. His Rysen system does not look like a sophisticated autonomous device at first sight. It is made of four cables hanging from two rails applied to the roof of the rehabilitation room and attached to a vest that the patient wears. The cables have just enough tension to help the patients stand and can be pulled or released to help them walk forward, left or right, without ever letting them fall. It brings a couple of key advantages over traditional rehabilitation methods: it provides a soft support, without unnaturally constraining the patient's joints as hard exoskeletons do, and it does not dictate walking pace or force the patient to walk on a straight line as treadmills do. The technological core of the system is the algorithm that decides how and when to pull the cables, and thus how much vertical and forward force to apply at each moment, helping patients to relearn walking, without having the machine actually walk for them. The algorithm must be personalized because no two people walk the same way and no two patients have the same lesion. In a study that made the cover of *Science Translational Medicine* in 2017, Courtine and his team studied how a group of healthy subjects used the robot.[7] First, they had them walk in the suspended vest with just enough tension in the cables to keep them strung and recorded the subjects' posture with cameras, their muscle activity with electromyography (a technique that measures electrical signals running through muscles), and the force applied by their feet on the ground. Then they began to pull the four cables up and forward in various ways, recording how the limb positions, muscular activity, and forces changed. They were essentially looking for a code, one that could tell them how a certain force applied to the cable translates into changes of the limb's behavior

After that, they did the same with impaired patients, asking them to wear the vest, applying just enough tension to keep them standing upright, and measuring how different each patient was from the kinematic, muscle activity, and forces that characterize "healthy" walking. Then they tried applying various configurations of pulls and releases in the cables to reduce that difference, searching for the configuration that allowed each individual patient to walk as much as possible to a healthy subject. It's not too different from what an optician does when deciding what kind of glasses you need: she tries various combinations of corrective lenses until you discriminate features as if you had perfect sight. At this point, Courtine's team could use the system to allow patients to perform walking exercises that they would otherwise be unable to do, and for much longer. The results, although tried on only five patients, were impressive enough: after a one-hour session, all of

Courtine's patients (who could all still walk, but only with some assistance) showed measurable improvements in speed and resistance without the robot, improvements that could not be achieved with a treadmill.

The next step was trying to help patients with more severe impairments. Patients who had lost the ability to control voluntary movements of the legs required additional help from another technology: epidural electrical stimulation (EES), which delivers electrical signals directly to the spinal cord after the lesioned region to replace signals that no longer arrive from the central brain. "Electrical stimulation has been used in patients with spinal cord lesions for about 30 years," Courtine explained when his 2018 *Nature* paper came out. "And in some cases, after intense rehabilitation, a few individuals were able to take a few steps, but only with the stimulation on. And until sometime ago, nobody understood why. So my team and I decided to look deeper into this question."

Courtine's team first studied healthy subjects to develop a detailed map of where the neurons in the spinal cord that control specific muscles in the legs are likely to be. Then they surgically implanted pulse generators in those spinal cord regions of three patients and fine-tuned the electrode's activity for each person. The team used magnetic resonance and computer tomography to select the electrode's configuration that best activated the relevant neurons of the spinal cord, and electro-encephalography to confirm that those neurons were also activating the motor areas of the brain's cortex—sending up signals about the limb's position in space. This way they were able to define, for each patient and for each movement, patterns of spatiotemporal EES that mimicked the instructions from the brain that would normally regulate the various phases of gait. Without voluntary contribution, the electrical stimulation induced minimal muscle contraction, but when the patient attempted to move their legs in sync with the electrical stimulation, it resulted in visible, although limited, movements of the hip, knee, and ankle.

At this point, the patients could start using Rysen to exercise on a treadmill. The robotic system took care of gravity, while the electrostimulation allowed them some voluntary movements of their legs. Patients had to synchronize their movements to the timing of electrical stimulation: their intentions actually translated into voluntary movements if they were in sync with the electrical stimulation. But that proved relatively easy to do, and in a few days, all three patients could enter a rehabilitation protocol. In the beginning, as soon as the EES was turned off, they lost all ability to move their legs' muscles. But after a few months of exercise, they regained control of voluntary movements even without EES. One of them could even take a few steps hand-free and without support after seven years in a wheelchair.

Courtine is now working to combine his system with Riener's Myosuit. The goal is to use both Rysen and the soft exosuit to improve rehabilitation and at the same time to improve Myosuit's control system so that it better complies with the natural gait of each individual patient. This would allow a seamless transition from the clinical environment to the patient's home when the gravity-assist robot is no longer required.

"The challenge for us is to go beyond a proof of concept and make this a treatment for everyone," Courtine explains. "And in this respect, we know that acting immediately after the lesion, when the potential for plasticity is still higher and the neuromuscular system has not yet degenerated, will be important. The key is for robotic systems to be transparent to the nervous system, to let it work without imposing their logic on it," Courtine says. "For this, the soft robotics trend will be crucial." Ideally, patients should not even realize they are wearing a robot, something that soft materials and actuators could make possible. "Also, in rehabilitation we must learn to not overassist patients; rather, we should provide as little assistance as needed for the nerves and muscles to do their work. For prosthetics, the situation may be different, because in some cases, you need to provide more assistive forces." Indeed, Courtine expects rehabilitation robots and prosthetics robots to diverge in the future. "Currently the same systems are used for both, but the needs are very different. In the future, it could be the same robots with different algorithms, but ideally they should be different devices."

In the meantime, as a future Nobel winner will probably acknowledge, other researchers around the world are working to help different groups of patients. Amputees who have lost their upper or lower limbs are also finding new options that were unthinkable only a few years ago. For decades, these patients could rely only on passive prostheses that restored the appearance of a limb but provided little or no functionality. In the late twentieth century,

the first electromyography (EMG)-controlled hand prosthesis appeared. Electrodes applied on the arm's skin could detect the residual activity from the stump's muscles and turn it into instructions for a robotic hand. But this technique could deliver only limited and unreliable signals to control gross movements such as opening and closing the hand. A big leap forward in precision came when EMG recordings from the skin were replaced by implanted electrodes connected directly to the muscle fibers, and things went even better when intraneural interfaces were perfected: these are electrodes attached directly to the nerves in the remaining part of the arm, thus connecting the brain to the artificial hand and allowing a more natural control. More realistic robotic hands, with independently articulated fingers and more degrees of freedom, became available, and their control became more natural for patients.[8]

In the mid-2010s, a team at Johns Hopkins University in Baltimore devised an ingenious and unprecedented solution for Les Baugh, a man who had lost both arms in an electrical accident forty years earlier. The Johns Hopkins team first conducted what they called "targeted muscle reinnervation," a surgical procedure to rewire to the muscles of the chest and shoulder nerves that were normally used to stimulate the muscles of arms and hands. Then they gave Les Baugh prosthetic limbs connected to his torso and shoulders and used the muscle activity of chest and shoulders to control the motors of the two robotic limbs. This way, the patient only had to think of moving his hand to actuate the robotic prosthesis. Baugh first practiced in a virtual reality environment and then started using his new arms with impressive success, even learning to grasp and hold small objects and to coordinate and control the two arms simultaneously, something his own doctors did not anticipate

Most prosthetic limbs still lack the sense of touch, though, and users rely on vision alone to make sure they are grasping an object the right way—not to mention recognizing its shape and texture. To address this problem, Silvestro Micera at Sant'Anna in Italy implanted tiny electrodes inside nerve bundles in amputees' upper arms and connected them to pressure sensors on the fingertips and palm of robotic prosthetic hands. Electrical signals from the robot pressure sensors were then fired directly into the nerves, providing patients with a sense of touch.

After this success, Micera used the same method to add proprioceptive information. He collected signals from tension sensors in the prosthetic hand and transmitted them to the central nervous system. Finally, relying on both position information and tactile feedback, amputees became able to manipulate objects more naturally with their robotic hands and were able to determine size and shape without looking.

Brain-computer interfaces (BCI) are a key technology to make robotic prosthetics more natural and effective. Pioneering work started in the late 1990s when Philip Kennedy and Roy Bakay in Atlanta implanted a locked-in patient (someone who is awake and conscious but unable to move or communicate verbally) with an electrostimulator that allowed him to control a computer cursor on a screen. In 2005, Matt Nagle, an American athlete who had been stabbed and left paralyzed from the neck down, became the first human to control an artificial hand via a ninety-six-electrode BCI developed by the American company Cybernetiks and implanted in his motor cortex. That device had been developed by John Donoghue, a professor of neuroscience at Brown University. While Nagle could only perform basic movements with the robotic hand, in 2012 Donoghue's group achieved another landmark result when two paraplegic patients could use their brain signals to pilot an external robotic arm and use it to grasp a bottle and lift it toward their mouth to drink. The result, published in *Nature*, made headlines around the world at the time.[9] Yet these were, by definition, very invasive devices that required complicated surgery, involved electrode boxes and bundles of wires sticking out of the patient's skull, and exposed the patient to risks of sepsis.

During the past decade, work has intensified on noninvasive BCI that uses electrical signals detected from electrodes glued to the scalp. However electroencephalography (EEG) is not as precise as implanted electrodes in discriminating the activity of neural groups. José Millán, first at EPFL and then at the University of Texas, realized that this problem could be circumvented if users learned to modulate their brainwaves so as to generate distinct brain patterns that the machine can pick up reliably and unambiguously. The development of training protocols based on more extensive data from users, together with machine learning techniques, made it possible to use distinctive brain patterns corresponding to mental tasks executed by subjects. By the late 2010s, the potential for EEGs was clear, not only for controlling robotic limbs, but also for better guiding rehabilitation. In a 2018 study on stroke patients, Millán's team used EEG signals to detect neural activity corresponding to hand extension and activate electrical stimulation of the corresponding muscles in the impaired hand, causing a full extension of wrist and fingers.[10] Patients using the BCI had a better and longer-lasting motor recovery than those receiving electrical stimulation alone.

Whatever the future holds for robotics applied to rehabilitation, prosthetics, and regeneration, success in this field will come down to one thing, as Conor Walsh himself told us: collaboration. No single technology will do the magic trick; only combinations of soft robotics, machine learning, electrostimulation, and neuroimaging, tailored to the specific conditions of each patient, will bring real breakthroughs. As Walsh says:

> The public often look at these devices and look for one person who is responsible for their development. But what is happening in the field is that people have learned to appreciate the importance of teamwork. If you want to make breakthroughs, you have to assemble the right team, with engineers, therapists, physiologists, neuroscientists—and you have to take the time to train that team, find a way to work together. I hope the mixing of disciplines we've seen over the last five, ten years will continue to grow. If it does grow, I expect to see more innovative designs and more products reach the market.

STOCKHOLM, SWEDEN, 2052

Over the 2030s and 2040s, several laboratories, including mine, reaped the fruits of the seeds planted by the pioneering work done in the 2010s and 2020s. Most important, the patients enjoyed the benefits. Ten years ago, my laboratory and the spin-off we created was the first to demonstrate, in a landmark clinical study of unprecedented size, the efficacy of a protocol that can be adapted to a variety of lesions and that in more than 60 percent of patients (including those with complete lesions) to regain almost all walking capability, provided that we can intervene in the very first days after the lesion. Dozens of other groups have obtained impressive results for both the lower and upper limbs. A diagnosis of spinal cord lesion or of paraplegia no longer means what it did thirty years ago.

As of 2019, half of the patients with spinal cord lesions remained paralyzed with life expectancies of decades with permanent disabilities.[11] Now the figure is down to 25 percent.

In the late 2010s, life expectancy in the United States for a spinal cord patient in his twenties was, on average, almost seven years shorter than the general population and had not improved since the 1980s. Now it is only two years shorter than for the rest of the population.

Thirty-five years ago, a complete paraplegic patient would spend on average thirty-five days in the clinic's rehab unit. Now the same patients can move to their homes, on average, after only twelve days—the time necessary to fit a device and refine a personalized rehabilitation protocol that then allows them to continue rehabilitation in their homes, guided by an app and weekly online sessions with a therapist.

A lot has also changed from those days with regard to the cultural attitude toward bionics, and the concept of a cyborg began to change. For many decades, the very idea of integrating artificial parts in a human body to the point of making it a human/machine hybrid was the stuff of science fiction. And it came across as a nightmare rather than a dream. Stelarc, an Australian performance artist, gained notoriety in the late twentieth century and early twenty-first by integrating robotics and neural control in his work. His whole work relied on the disturbing effect caused in audiences by the visible, often violent, intrusion of technology into biology—like when he let the audience control his body by activating muscle electrical stimulators over the Internet.

Things began to change when popular figures such as paralympic athlete Aimee Mullins, former car racer Alex Zanardi, and Hugh Herr did nothing to conceal their status as amputees. The increasingly natural appearance and compliance of limb prostheses, the miniaturization of brain-machine interfaces, and the recent machine learning algorithms that rapidly personalize and make the control effort disappear, did the rest, to the point that it is now socially acceptable to present yourself as a human/machine hybrid.

The increased acceptance of robotics body parts meant less stigma for patients, but things quickly went beyond that. Healthy people have begun using robotic garments to improve strength and reduce fatigue when painting their roof, moving furniture up and down stairs, and to have more fun when hiking, skiing, or sightseeing. Cities now brim with indefatigable "augmented" tourists who manage to walk to twice the sightseeing spots they would once have visited.

This has created new problems, from fairness in sports to totally new dangers. Many people now have accidents for using these devices the wrong way, ironically often ending up with the same impairments that exosuits and prostheses were created to cure. There are new sources of inequality between who can afford these technologies and who cannot. As it is often the case, when science and technology overcome old challenges, they end up creating new ones. But we can all be proud of what the sector has accomplished and of how we have improved the quality of life of so many patients.

8 MICROSURGEONS' FANTASTIC VOYAGE

Thomas was, understandably, a little nervous. As he was wheeled into an operating room at Kigali's main hospital, he was trying to remember the exact wording of the risks and possible side effects listed in the consent form he had just signed. "I don't need to worry, do I?" he asked the nurse, forcing himself to smile.

Ada, the nurse, had a long experience of comforting and reassuring patients before surgery. But she was still perfecting the right words to describe this particular procedure to patients, giving all the right information and presenting the advantages and risks in the most honest way.

"You are about to be treated by one of the best teams in the world," she began. "And you are lucky to be living in this time. Not too long ago, patients in your condition would just have to wait, hope for the best, and make sure to never get too far from a hospital. Now we can treat and send you home by this evening."

She had a point. Thomas had a brain aneurysm, a bulge in the blood vessels of his brain that had probably been there for some time and had been discovered during a routine check a couple of months earlier. Looking at the MRI scan, a neurologist had determined that there was a significant risk that the aneurysm could rupture within one or two years, possibly causing extensive brain damage or death. Up to the previous decade, there were not many options for patients in this situation, especially for a seventy-one-year old patient like Thomas. An invasive brain operation that involved removing a piece of the skull to access the brain and clamp the aneurysm was an option that surgeons considered only when an aneurysm looked *really* scary and was sending warning signs of being about to rupture in a matter of days. And it was an option they would typically contemplate for younger patients, because for someone who was Thomas's age, the risks of the procedure would almost equal the risk posed by the aneurysm itself. Most doctors, back in the day, would prefer to monitor brain aneurysms and avoid surgery until the rupture happened.[1] But now there was a much safer option to get into the brain's vessels and get rid of the aneurysm. "No procedure is without risk," Ada told Thomas. "But I can assure you that unlike it was in the past, the risk is far, far lower than the risk of doing nothing."

Ada was right, but the technology was still novel enough to instill doubts in some patients who had reservations at the idea of having those things going around their body. And it was not exactly a routine procedure yet.

Meanwhile, in another room, the team of surgeons who were about to carry out the procedure were going over their checklist once again and mentally rehearsing it all. First, as the chief surgeon was explaining, they would inject a handful of micron-scale robots through a catheter in a leg artery. Then they would guide them with electromagnetic fields along the patient's body and through the blood vessels, all the way to the location of the aneurysm. Once there, they would signal to the microrobots to start clipping the aneurysm.

"I still can't believe that we are able to do this," one of the assistants said.

"I have a hard time believing it myself, and I've done it quite a few times by now," the chief surgeon replied with a smile. She had been among the first American specialists to perform the procedure on actual patients, after long training sessions on robotic models of the human body and then on animal models. "When I entered medical school, this still sounded like a crazy idea out of a science-fiction movie," she recalled. Building robots of that size, making them biocompatible so that they could safely stay in the organism, and making them functional enough to execute a procedure on a brain vessel sounded crazy enough. But controlling their movement from outside without a tether? "I can tell you most people did not believe we would ever get there," she recalled. "And instead, here we are. In a couple of hours we will have cured a brain aneurysm without leaving a single scar on the patient."

At that moment, Ada's voice came through the speaker. "Good morning, doctors. The patient is ready for anesthesia. Would you like to meet him before we proceed?"

"Sure we do," the surgeon said, signaling to her team to quickly finish their preparation and move toward the operating room.

"Good afternoon, Thomas," the surgeon said. "How do you feel today?"

"Good morning, doctor," the patient replied, looking up toward the big camera-equipped screen hanging from the wall. "I feel well, but I'll feel better when this will be over. I'm sorry you had to get up so early this morning. What's the time there? Six?"

"No problem at all, Thomas," the doctor replied. "Yes, it's six a.m., and it's not even our first procedure this morning. We have to do it this way sometimes, but it's still much easier for us than flying over there or having you fly all the way to Boston."

It wasn't lost on the patient that not so long ago, that level of health care would have never been available for him. Upper-middle-class Rwandans like him could surely rely on good doctors and fine hospitals, but high-tech surgery executed by one of the best specialists in the world was another matter. At the very least, even patients who could afford the procedure would need to fly to the United States or a few other countries for treatment. But the massive robotization of surgery in the previous decade had changed that, turning many procedures into yet another thing that could be done remotely, like work meetings and yoga classes. So there he was now, speaking via teleconference to an A-list surgeon who, on the other side of the world, would soon be guiding an army of minuscule robots through his blood vessels and to his brain.

"I am sure Ada has already explained everything to you in detail, Thomas," she told him. "In a few minutes, after anesthesia, she will use a very small catheter to inject a few nanoagents inside your body. They are not toxic; in fact, they are completely biodegradable and your body will get rid of them a couple of days after the procedure. After the injection, I will remotely control a magnet there in your room and guide them toward

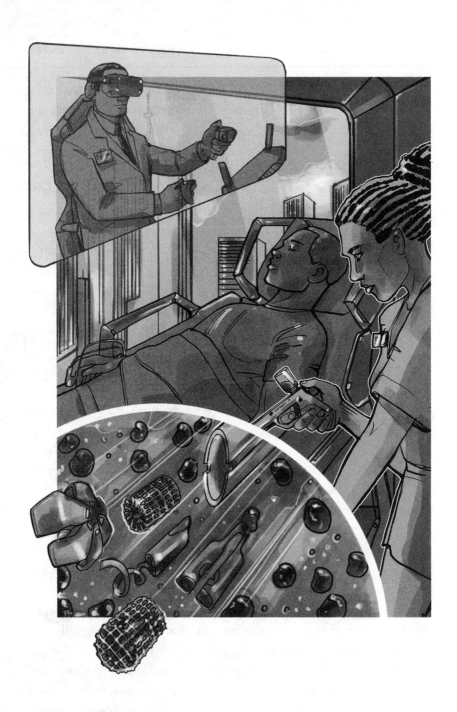

your aneurysm. You see that small screen on the side of the bed? That's for Ada, but I will be getting exactly the same images here, tracking the robots as they move toward your brain. Once they are at the aneurysm, I will use acoustic signals to instruct them to release tiny metal plates that will fill up the aneurysm, cutting off its blood flow, and making sure it will not break and cause a hemorrhage."

"That sounds like a plan, doctor," Thomas replied, trying to sound a little bit calmer than he was as Ada began to administer the anesthesia.

ZURICH, SWITZERLAND, AND PISA, ITALY, CURRENT DAY

Of all the stories in this book, this may be the one that most closely resembles a classic work of science fiction. For fans of 1960s cinema, the idea of an army of "microsurgeons" entering a person's body and traveling through the veins up to the brain may sound familiar. In the Oscar-winning 1966 movie *Fantastic Voyage*, a crew is shrunk to microscopic size thanks to a futuristic physics-altering technology and injected in the body of a comatose scientist whose survival is critical for the future of humanity. The crew has one hour to remove a blood clot from the scientist's brain, something no ordinary surgery could do.

That movie and our fictional story share a very real premise: there are a number of medical procedures—not only those involving the brain—that are still difficult and risky for normal surgery and sometimes outright impossible. Technology has made things better in many fields, with laparoscopy and "classic" robotic surgery, which uses finely controlled microinstruments that get into the body through small holes instead of several large incisions. But things get more difficult when the organ to be cured is behind a layer of bone tissues—as in lung surgery and, most critically, brain surgery, or when the problem is in the tiniest vessels of the circulatory system.

Shrinking humans to microscopic size would be great, but it can happen only in movies with the help of special effects. Injecting tiny robots in the body to deliver therapies instead is far less fantastic than it may seem. Engineers have been studying the problem for many years, experimenting with a wide range of materials, designs, and control strategies in order to get microscopic robots into the body. And though none of these experiments has made it to the clinic yet, they proved that the technology works and that microrobots could one day be used to deliver drugs to the precise locations where they are needed, to operate microscopic grippers or cutting devices for surgery, and to replace DNA within a cell, to mention a few.

Born and raised in the Midwest United States and a professor at ETH Zurich since the early 2000s, Brad Nelson has published hundreds of scientific articles—and received patents—describing the fabrication and control of microrobots. Along the years, he developed a variety of micromachines, such as microrobots to inject DNA into cells;[2] tiny devices to automate cloning and genetic manipulation while reducing contamination; microdevices controlled by ultrasound for grasping and aligning cells suspended in water;[3] microrobots with bacterial flagella that swim in fluids and manipulate small particles;[4] and many more variants of the *Fantastic Voyage* concept, experimenting with diverse materials, fabrication techniques, and modes of locomotion, from micro (less than 1 millimeter in size) to nano (less than 1 micron, thus 1,000 times smaller) robots.

At micro- and nanoscales, Nelson notes, robotics becomes almost a different discipline from its macroscale version. "When you take something and shrink its size by a factor of 10, its area goes down by a factor of 100, but its volume reduces by a factor of 1,000. Volumetric properties become much less important, and surface properties start to dominate," he notes. "You have to start thinking about surface interactions in a very different way. Roboticists typically see interactions between objects in terms of the Coulomb friction (a model of how dry surfaces exert friction between each other), which is what lets you hold a glass in your hand, but that is no longer so important at these scales. But we can still use a lot of the same algorithms, for example, for sensor fusion and navigation. When we're trying to get a microrobot to get through blood vessels, the algorithms we use are similar to the ones we use for getting a car through a city."

Building the robot itself is not the biggest challenge in medical microrobotics. We are, after all, in the era of nanotechnology, and a number of tried and tested techniques exist to manipulate all sorts of materials at those scales. For example, Nelson and several other researchers have built artificial microstructures made out of organic materials such as sugars, lipids, gelatins, and inorganic ones such as manganese, zinc, and other metals.[5]

The problems begin when the robots must move inside the body in a controlled fashion. "Especially in vessels, you have to deal with blood flow. There you have surface interactions that are very difficult to model and require in vivo studies—which means you need animal trials or human clinical trials to understand what happens," Nelson notes. A number of strategies have been tried to power the robots through the body, including equipping the microrobots with their own reservoir of fuel that can react with the external environment and using the energy from this reaction to propel the robot.

But the most promising strategy is to use magnetic fields to guide micro-robots.[6] By placing tiny magnetic beads on the head of a microrobot and applying a low-intensity magnetic field from the outside (which is harmless for humans), it becomes possible to drag and guide the robot toward its final destination in the body. Or if the robot has a helical tail and the magnetic field rotates, the robot will rotate like a propeller and move even faster.

Magnetic actuation is not without challenges, though—especially if it is used for drug delivery, one of the most coveted applications of medical microrobots. "It's a physics textbook problem," says Arianna Menciassi, from Scuola Superiore Sant'Anna in Pisa, Italy. "The force that attracts a magnetic object toward the magnet is the result of the product of their volumes. If one of the two volumes is extremely small, as in this case, the other one must be quite big. But there's a limit to how big a magnet you can have in a clinical setting." That means the magnetic field will typically be weaker than what would be needed to drive all microrobots precisely toward their destination. And that can be a problem, in particular for drug delivery, where the idea is to inject a large number of drug-loaded robots and drive them magnetically toward the specific location in the body. If too many robots are left behind because the magnetic field does not pick them up, they will accumulate and deliver drugs to other parts of the body. In that case, the main selling point of microrobotic therapy—reducing unwanted side effects—will vanish. "Either you are very good at accumulating robots where you want them, or you have to be very good at recovering those that go unused," Menciassi explains.

Menciassi and her team have recently proposed to go down the latter road and associate drug-loaded "nanoagents" to a slightly more traditional macroscopic technology: magnetic vascular catheters that can get to an organ through a vein or artery.[7] They demonstrated a system where one such catheter could get inside the kidney through an artery and inject a large number of such drug-carrying nanoagents. The tiny robots would then be attracted to the target site (such as cancer tissue.) by either an external magnetic field or biochemical forces such as those that bind a receptor to a ligand. After a while, another catheter, this one coming to the organ through a vein, would act like a magnetic vacuum cleaner, retrieving all the nanoparticles that did not get to the tumor and removing them from the bloodstream.

Another big challenge is seeing what the robots do, Menciassi explains. Microinvasive—or noninvasive—versions of surgery are a great idea, but doctors will still want to see what is going on inside the patient's body. "We explore the combination of ultrasound and magnetic fields to track the robots," says Menciassi. "The idea is to use the magnetic field to cause the microrobots to vibrate at a certain frequency, and then measure the ultrasound signals they produce to produce images of their movements inside the body." The method relies on analyzing the tiny, high-frequency sounds produced by the vibrating microrobots and measuring the Doppler effect— the shift in frequency that you hear when the source of a sound is moving toward or away from you.[8]

All these techniques work on microrobots made out of metals, carbon structures, or organic compounds. Sometimes the shape and movement of these structures are inspired by biology—for example, when they are powered by tails or cilia that make them swim much like bacteria. But they are still 100 percent artificial. Some researchers are trying to use biology itself, rather than bioinspiration, turning bacteria or other microorganisms into hybrid microrobots.[9] For example, if the goal is to deliver drugs at a precise location in the body, the drug can be placed in a micro "backpack" attached to a bacterium like *Escherichia coli*, and a magnetic bead can be attached to its head. This way, the magnetic field can be used only for steering while leaving the swimming task to the bacterium's exceptional abilities. Or, even better, researchers can use magnetotactic bacteria, which contain magnetic crystals and are naturally capable of following the orientation of the magnetic field, an ability that in nature they seem to use to move toward their preferred concentration of oxygen.

Menciassi jokingly says that the difference between the artificial and biohybrid approaches is analogous to choosing between "making an artificial cat or training your own." The latter option sounds much easier, one has to admit. Bacteria are already very good—often *too* good—at moving around our bodies. One would only need to teach them where to go. Of course, before using them as microdoctors, you would need to tweak their biology a bit so that they do not release toxins or multiply uncontrollably and cause an infection, but that is not impossible for modern genetic engineering.

Still, Menciassi thinks that getting these hybrid robots into clinical trials and authorized may take longer than for artificial microrobots. "Magnetically guided microrobots will probably be the first technology to get to the clinic," she says, "not least because there is already magnetic equipment in most clinics that can be adapted, such as magnetic resonance devices."

As if building and controlling a mobile microrobot was not challenging enough on its own, some researchers are trying to learn to use dozens, even hundreds, of them at the same time, steering them into collective and choreographed behaviors. At the Max Planck Institute for Intelligent Systems in Stuttgart, Germany, Metin Sitti brings together microrobotics and swarm robotics. When asked why he is trying to solve two different problems at once instead of focusing on just one (swarm robotics is difficult to do at any scale), he says that in many cases, a single microrobot may not be capable of doing the work that medical applications require. "A single microrobot can deliver only tiny amounts of a drug, or apply very little heat to an area. Most functions, such as delivering drugs or removing cancer tissue, require cargo capacity, or coverage of large areas." But if many small robots join forces after having been injected, they can approach the functionality of a macroscale device and still enter the body in a noninvasive way. "Equally important, a single microrobot is very difficult to see," Sitti explains. "How do you image and track a robot with the size of a cell inside the body from outside? A robot swarm is easier to detect from outside as long as you manage to program and control it."

Inducing and controlling the swarming behavior is the big challenge here, and the best way to achieve it depends on what type of robot one starts with. "For remotely actuated robots, ultrasound is becoming increasingly interesting for controlling microrobots," Sitti explains. "Ultrasound waves allow you to get large numbers of particles to assemble in different configurations, and you can easily change the wave patterns to induce different behaviors. But because the external signal is going to be the same one for all robots, you cannot really have individual control of each robot." How can you create robots with specialized behaviors then? One of the possible solutions is to program and control the dynamic or static interactions between the microrobots by remote acoustic or magnetic fields to induce different programmable self-organized patterns and dynamics.

In a 2019 article, for example, Sitti showed how to get a set of passive particles to self-assemble into a four-wheeled microcar and start moving once an external magnetic field is applied.[10] To do this, he and his team leveraged the fact that at this scale, the shape of the robot's external surface alters the strength of the electric field around it. Fillets or cavities, for example, create different—and predictable—magnetic fields that in turn have predictable effects on other magnetic tiny bodies. Sitti's microcar is made of a rectangular body with four semicircular cavities at its corners and four tiny spheres of magnetic material. When a magnetic field is applied, it attracts the spheres toward the cavities and makes them rotate in the direction of the magnetic field, to the point that they become wheels that propel the car in the desired direction. Self-assembly and locomotion arise without the need of controlling the individual particles, but rather from careful design so that the magnetic fields generate appropriate attractive and repulsive interacting forces among the particles. The same logic can be applied in different interactions—chemical, capillary interactions, fluidic forces—to obtain different shapes and different forms of locomotion or actuation. "You can do this with magnetic fields or with acoustic signals—using different notes to change the dynamic or static equilibrium and create different programmed patterns," he explains.

Things are a bit different with self-propelled hybrid or chemical microrobots, Sitti says: "That becomes a bit more similar to swarming at the macroscopic level, because you can rely on the intelligence and locomotion ability of the individual units, but you may still use external stimuli to steer and control the swarm." For example, bacteria-based robots may tend to follow chemical signals or oxygen gradients, whereas algae-based ones are very sensitive to light.

"Magnetic control is very popular right now and has been studied more than other ways to control microrobots, but acoustic is also very interesting," Sitti continues. "It is a long-distance force that can access more depth than the magnetic field and is very strong even at small scales, while the magnetic force scales down with the robot's magnetic volume. The signal can be much more diverse, you can have specific patterns encoded in the waveform, and you can program it easily." The latter method allows the creation of more complex choreographies, where the microswarm changes configuration and performs different actions over time in response to the "melody" played by the acoustic field in ways that would be difficult to achieve with a magnetic field.

Magnetic and acoustic methods can be combined, as Sitti showed in 2020 in a proof-of-concept study where he and his collaborators built 25-micrometer-long, bullet-shaped microrobots with a spherical cavity.[11] Once immersed in a liquid, the air bubble inside the cavity can be made to oscillate by bombarding the liquid with acoustic waves at a frequency that puts them in motion—surprisingly fast motion of up to ninety body lengths per second. A fin on the back of the bullet helps maintain a forward direction, while a magnetic coating enables precise steering by means of external magnetic fields. The propulsion force of these microrobots is two to three orders of magnitude stronger than that of algae and bacteria, more than enough to move them even against the blood flow in the vascular system and inside the highly viscous and heterogeneous mucus layer of the gastrointestinal tract for future medical applications.

Nelson agrees. "We are already seeing magnetic strategies being used to guide catheters through the body," he notes, "and in the near future we will see smaller catheters, below 1 nanometer in diameter. Untethered devices are going to be further away, but in five to eight years, I think we will see them moving through the body, at least at an experimental level. It takes a while to understand the problem. But as with most things in engineering, once somebody gets there, everyone will think, 'Of course you did it. What's the big deal?'"

Micro- and nanorobots are good demonstrations that "there is plenty of room at the bottom," as physicist Richard Feynman famously said in the early 1960s envisioning the potential of nanotechnology.[12] In the case of medical robotics, there is also plenty of room before reaching the actual bottom. Miniaturized—but still macroscale—devices can offer alternatives to invasive methods and be guided by magnetic or other external fields to perform surgery inside a body. They just need a different door to enter the body. Injecting anything larger than a few micrometers would be a terrible idea (it would clot blood vessels), but swallowing it is perfectly fine.

Daniela Rus's group at MIT developed an origami robot folded within an ice pill that can be simply swallowed. As the ice pill melted, the robot unfolded inside the stomach, rolled toward a button battery, and removed it from the wall of the intestine. Now, if you think that removing batteries from a patient's stomach is not a problem that deserves too much attention, you are very wrong. More than thirty-five hundred people of all ages ingest button batteries in the United States every year.[13] Once they've been ingested, there are not many options to get them out in time: over the past few decades, some dozen deaths happened due to complications, mostly among children who mistake button batteries for candies.

In their initial 2016 study, Rus and her team designed two slightly different origami robots: one for removing the battery and one for delivering drugs to cure the wound caused by the battery to the stomach wall.[14] The little "remover" is a cubic magnet, designed to work from inside the ice capsule as it melts. The idea is that the patient swallows the ice capsule with water, and an external magnetic field is used to guide it to the stomach and close to the battery, which is attracted by the magnet in the ice capsule. At that point, the tiny robot and battery are guided through the gastrointestinal tract and expelled. Then the "deliverer" would get to work; it's an accordion-shaped origami folded inside the ice capsule that expands by five times when the ice melts. The deliverer includes a drug load and a small magnet that make the unfolded robot contract and expand toward the desired location under the effect of the external magnetic field. The robot is then deformed by the magnetic field to patch the ulcer caused by the battery and medicate with the drugs.[15]

In order to have at least a proof of concept that the system may work, Rus and her team had to develop an artificial model of a stomach. "We needed a simulated organ that preserved some of the key properties of the physical one," she explains. The scientists used 3D printing for the general shape and then molded the inside of the stomach to obtain a soft, tissue-like material inside. She continues, "The approach I always like to take in robotics is to start with conceptualization, then develop the mathematics, then start implementation and testing. For testing, we typically start from simulation, which is extremely useful but has a big gap with reality. One way to close the gap is to proceed in steps. In the case of the pill, the first step was working out the control system in simulation. Then we built this simulated organ and actually moved the ice pills and the battery through it. The next step will be a more realistic environment—ideally, the stomach of pigs."

Rus says she would like to see her idea reach clinical settings in five to ten years. "First, we would need funding for in vivo tests, which will take some years. Then there are at least a couple of years of human trials, and you get to ten years very easily. But it would be terrific to have this technology and be able to move inside the body without having to cut. As a field, we have already contributed a lot to medicine: the fact you can see inside the body without cutting is due to computer imaging. Our work represents the next step, acting inside the body."

Indeed, mini- and microrobots, like all other medical devices, will have to be thoroughly tested and approved by regulators such as the Food and Drug Administration in the United States or the European Medicines Agency in the European Union. Nelson notes that using soft robots, and soft microrobots in particular, can compare favorably in a risk analysis to rigid tools such as catheters. "You can't puncture things, you can't injure the patient, so they have to be safer—provided you can show that there is an actual benefit and that you can bring an improvement over existing therapies." That is why, he notes, it is crucial to find the right clinical application: "I think that vascular diseases could be among the first. Things like strokes, aneurysms, occluded arteries. Gastrointestinal diseases are also a promising area, as well as removal of emboli and tumor ablation, a procedure that burns or freezes tumor cells instead of removing them, and is currently performed with needles."

In the end, Nelson sees microrobotics as contributing to what was from the start one of the main selling points for robots in medicine: democratizing surgery. "Right now," he notes, "there are procedures for which surgeons have to train for years. Laparoscopic surgery, for example, is difficult. The promise of this field is to make some procedures easier to learn so that less experienced surgeons can do more difficult and expensive procedures." The other promise is telemedicine. "That was really the original idea of robotic surgery, as it came from the military who wanted a way to treat soldiers in the field right after they're injured, without the doctor standing in harm's way. Medical robotics came from the US military, and later the idea was picked up by companies such as Intuitive Surgical," he says, referring to the maker of Da Vinci, the world's leading surgical robot. The idea has become even more important after the COVID-19 pandemic. Having surgery performed by tiny robots inside the body instead of the surgeon's hands would make physical proximity between the doctor and the patient even less crucial. "The surgeon doing the operation does not have to be in contact with the patient," he says. "You can have the patient in one place and have the best specialist in the world operating from another place, making lifesaving therapies more widely available."

KIGALI, RWANDA, MAY 12, 2038, TWO HOURS LATER

"Welcome back, Thomas. How are you feeling?"

Ada's smile was the first thing he saw after opening his eyes and quickly adjusting to the neon light pointing at him from the ceiling.

"A little dizzy, but good," he replied. "How did it go?"

"It went perfectly. The aneurysm is no longer a problem, although we will keep it in check for some time. Get some more rest. We'll keep you under observation for a couple of hours just to be on the safe side, and then you'll go home."

"I wanted to thank the surgeons. Are they still online?" he asked.

"I'm afraid they aren't," Ada replied. "They're probably already in the middle of another operation—in Russia, I think."

Thomas lay his head back on the pillow, took a long breath and smiled, thinking that he would be home for dinner that night. Then a thought crossed his mind.

"Are they still in my head?" he asked Ada.

"The nanorobot, you mean? Yes, they will be for a few more hours, and then they will be gone, more or less like the coating of a pill. When the magnetic field is turned off, they are just transported in the blood and slowly dissolve. But it's better to remain here with us a couple more hours in the very unlikely event of an immune reaction. By the way, we have a few things to help you pass the time. Would you like some music or a movie?"

"A movie would be great," he replied with a smile, browsing through the list of titles that the touch screen next to the bed was showing. "But definitely not *Fantastic Voyage*! I've had my own version already today."

9 LIFE AS IT COULD BE

HUNAN PROVINCE, CHINA, 2051

Entering the restaurant was a bit of a shock. They had been outdoors since the early morning, feeling all the time as if they were inside a National Geographic documentary. For hours, they had heard only birds sing, monkeys screech, bigger animals growl. Now the crowded, noisy, and air-conditioned room was a rough return to reality. And this was the luxury restaurant of the park, renowned and celebrated by dining magazines for the pioneering work of its chef. "If this is so crowded, who knows what the cafeteria downstairs must be," said Javier, taking out his backpack and signaling to the maître d'. Luckily, their host had reserved a quiet table for them in the back of the restaurant, where they were quickly seated. They all crashed on the seats around the table, at the same time exhausted and excited for what they'd just seen.

"Wow!" Corinne began. "That was definitely worth waking up at four in the morning. I still can't believe what we've seen, guys."

"I keep thinking about that tiger catching that boar right in front of us," said Javier, recalling a particularly spectacular and crude moment, when they were observing two tigers hunting from a bridge passing over the big cats area.

"Yeah, that was tough to see," replied Corinne. "I know I should not feel anything for the boar, but it was hard not to."

"And what about the bears catching salmon as they jumped up the falls?" Katsuo chimed in. "I've watched the scene so many times in documentaries, it was crazy to see it so close up."

"For me the highlight was the encounter between the alligator and the anaconda," said Aya. "At some point, I really had problems telling which of the two was the real thing. You know, anacondas can prey on alligators sometimes, so it could go both ways in principle."

"Maybe they were both real, or both fake," said Corinne. "We assumed only prey could be fake, but who knows?"

"Okay, here comes the person who can answer all our questions," said Javier.

An elegant and energetic woman in her forties had just entered the restaurant and was walking toward them with a smile, casually responding to the servers' respectful salutes along the way.

"Hi, guys. I'm Mei. Sorry for being late," she said, as she quickly shook everyone's hand before sitting on the empty chair that had been waiting for her at the table. "Have you ordered already? Oh, you didn't have to wait for me."

"It's not a problem. We know how busy you must be."

"Oh my goodness, it's even later than I thought," she said, looking at the clock on the restaurant wall. "I'm so sorry. I should know better than trusting this watch."

"Is that the GenWatch 4?" asked Katsuo in amazement.

The GenWatch was all the rage among sustainability-minded people like Mei. Instead of quartz and batteries, it kept time thanks to a population of bacteria, genetically engineered so that the molecular machinery of their circadian rhythms could count seconds and minutes—but not always accurately.

"It is better than the previous one, but there's still no way it can beat a traditional watch," Mei said. "But at least it will never become electronic waste. And neither will all the things you've just seen in the park. So what do you think of it?"

They didn't even try to hide how excited they were and flooded her with compliments that she received with a quiet smile, as if it was something she was used to.

After the excitement came the questions. How many prey did the animals consume each month? Which ones were more difficult to build? How many were remotely controlled, and how many fully autonomous? She took note of a few questions as if she were a speaker at a conference.

"So, let me start from the beginning," she said calmly. "As you may remember, when we first created the park, it was at the same time a scientific experiment and a conservation project. We wanted to save some animal species from extinction and at the same time study their behavior in a much more realistic and controlled way than what had been done up to that point."

"Realistic," she explained, because the animals would be able to hunt, chase, and capture prey as if they were in their natural environment instead of being fed raw meat as in normal zoos. And "controlled" because, by tuning the behavior of those prey and of other species that were parts of the ecosystem, researchers could study the ethology of the animals—from how they hunted in groups to what made a particular prey desirable.

The key to making the park possible, she explained, was that by that time, scientists had mastered not one but two huge challenges that still looked daunting a few decades earlier in the 2010s. On the one hand, they could make robots that looked like real animals and moved realistically too by using soft materials for skins and joints and machine learning algorithms to govern movements. On the other side, they could make those robots out of biodegradable and digestible materials that living organisms—from bacteria and mushrooms to mammals and even humans—could consume. In other words, they could not only create robotic versions of mice, fish, and other simple animals that were realistic enough to fool their predators. But they could make those robots edible, so that the predators could chew and digest them.

"We sort of piggybacked on decades of research on biodegradable robotics, which of course was not done with a theme park in mind," Mei acknowledged. In fact, concerns about the environmental impact of electronic equipment began to grow in the late twentieth and early twenty-first centuries as computers and smartphones were sold (and disposed of) in the millions. When the prospect of a robotic revolution—that would have turned robots into mass products and flooded the world with them—started to look closer and closer, many researchers became aware of the contradiction. Robots were supposed to contribute to a more sustainable world, but at the same time they risked contributing to pollution from electronic waste, to the unsustainable extraction of rare earth materials used in semiconductors or batteries, to energy consumption.

At the time, soft robotics was gaining popularity among researchers as a way to make robots more versatile and more adaptable to different environments. To roboticists, the overall appeal of soft materials lay in how their functionality—the way they could bend, adapt to objects, change shape—could approximate the behavior of living tissue in a way that hard materials could not. Why not take the concept to its logical conclusion? Why not use materials that were based on the same organic component that make up living tissues and could degrade in the environment at the end of their lives? In addition to the environmental concerns, some of these researchers also had clinical goals: to create devices that could enter the human body—for example, to deliver drugs or perform diagnostics—and then be metabolized instead of having to be taken out.

Starting from the 2010s, Mei recalled, scientists began to experiment with making robotic components from biodegradable and digestible materials. It was a long journey from those first edible actuators and electronic components to what the group had just witnessed in the park. But many failed attempts and false starts later, a group in China managed to create a small, walking robot entirely made of organic and edible materials that would decompose once switched off. They then managed to make it resemble a mouse, to the point that reptiles and mammals would be fooled into hunting it. Then they did the same trick with fish.

"At that point, they reached out to us, and we immediately started collaborating," Mei recalled. "At the time, this place was a natural reserve. We had a few South China tigers here, whose population in the wild was down to a few dozen animals because their habitat has been disrupted by urbanization and because they struggle to find prey in the wild."

The problem, she explained, was that tigers had difficulties in finding prey in the reserve as well. It was impossible to recreate their entire ecosystem there, no matter how much the team tried. "We would feed them raw meat, but that made the animals the ghosts of themselves. They got fat, lazy, and depressed. So we asked those researchers to make us some soft robot rodents that tigers could hunt and eat. And after a few attempts, it worked."

From then on, Mai explained, it was all downhill. If it worked for tigers, it could work for wolves, snakes, alligators. It could work for bears that feed on insects and larvae, as well as for those with more refined taste. "We realized that scenes of wild hunting could make our park more attractive to tourists, and we began to take some liberties, such as the bears hunting salmon, which are not really from this region but visitors apparently are crazy about. I mean, who doesn't love to see the bears catching salmon midair?" Katsuo blushed but said nothing. Mai added that some animals were still getting meat for dietary reasons, mostly lab-grown meat cultured in bioreactors, the kind of meat that was becoming a staple of most restaurants for humans too.

Yet there was also another side to the park, less noticed and less attractive for tourists, but not less important. It was a place where hundreds of biodegradable robots and electronics components from all over China were discharged at the end of their life cycle, in parts of the park hidden to visitors, to be consumed by bacteria, mushrooms, and insects. In part they just fed the ecosystem of the park, helping the growth of flowers and trees that were then moved and transplanted throughout the park area. And in part they were an energy source, thanks to microbial fuel cells that extracted electric energy from the decomposition of soft robots.

"That energy feeds a lot of what happens in the park," Mei said. "Speaking of feeding, I think it's time to order. What do you think?"

Everyone nodded vigorously. Captivated by Mei's recount of a few decades of research into edible robots, they had almost forgotten how hungry they were. Mei looked toward the server, who immediately came to the table and addressed the guests with a well-rehearsed presentation of the restaurant's special testing menu.

"It's a pleasure to have you all here," he began. "We have some unique specialties on our menu today. I hope you will like them, and I am already quite sure they will surprise you."

Today's robots are mostly made of plastics, metal alloys, and composites with specific properties dictated by function. These materials cannot be found in nature. They are fabricated with processes developed over thousands of years: high-temperature molding of the copper, bronze, and iron ages; blending and controlled cooling of the steel age (industrial revolution); and automated composite processing of the current age of plastic and silicon semiconductors. However, the fabrication and disposal of these materials have a negative impact on the environment and our health: they exhaust natural resources, consume energy, and pollute air, soil, and water. Electronic waste has been recognized as the fastest-growing type of hazardous solid waste in the world.[1] While the production and disposal of robotic technologies still represent only a minimal fraction of the broader information technology industry, roboticists have already started to investigate the use of safer and healthier biological materials.[2]

There are two key differences between biological materials and engineered materials: biological materials are made of a smaller number of substances and self-assemble through growth, whereas engineered materials can leverage a substantially larger number of elements and are fabricated according to a precise plan.[3] The properties of biological materials are mostly due to the intertwining of microstructure and shape resulting from the growth process, whereas the properties of engineered materials are due to the selection and processing of the constituent elements.

For example, plants move their organs by using many different mechanisms, even when they are dead. These movements are mostly generated by modifications of the material microstructure induced by humidity, temperature, and light.[4] The chemical process of osmosis—the transfer of fluid across a membrane from a less concentrated solution to a higher contracted solution—is often leveraged by plants to generate motion. The walls of plant cells are made of stiff cellulose fibers embedded in a soft matrix that can swell perpendicular to the direction of the fiber by osmosis. The superposition of multiple layers of swellable cells with different fiber orientations generates bending and torsion. For example, pine cones open and close at different times of the day as a result of the bending of their scales under changing humidity conditions. While these movements are relatively slow compared to those generated by animal muscles, some plants can be very fast, such as the rapid closing of the flower of *Dionaea muscipula* (Venus flytrap) or the sudden folding of the leaves of *Mimosa pudica*. These rapid movements are produced by cells on the surface of the plant organ that translate mechanical stimulation into electrical signals, which rapidly change the permeability of cells in the structure, thus resulting in deformation of the plant organ. Engineers have taken inspiration from these mechanisms to develop osmotic actuators triggered by electrical pulses or made of composite materials that bend by absorbing and releasing vapor.[5] But these bioinspired actuators are not biodegradable and are designed for predefined and immutable operation, unlike plant materials that can dissolve and provide nutrients, and while they are alive, they can also adapt to changing conditions and self-heal through growth.

In order to transition toward biodegradable and even edible robots, researchers are rethinking the entire process of robot design and function, and they start directly from biomaterials. For example, at Tufts University in Massachusetts, Fiorenzo Omenetto has a lab that is almost entirely devoted to working with silk. This organic material has toughness and mechanical properties that are superior to the best manufactured fibers such as Kevlar, and Omenetto has shown that it can be used for a wide variety of applications, from biodegradable cups to electronics and light sensors. He collects the material from silkworms, then processes and mixes it with additives instead of recreating it synthetically. He says:

> Ultimately it is a thermodynamic problem. The moment you break down a material like silk into its basic components in order to analyze it, you have to add energy to the system. Proteins get denatured and lose their properties, and you don't have a way to reverse the process and go back to the original material. There is a lot of interesting research in making artificial silk, but it has not delivered results comparable to natural silk. It is extremely difficult to imitate all the properties of the material. It is not a complex structure in principle, but the devil is in the details. If you change just a few of them, such

as the exact length of the protein or the positions of binding domains, you lose the very properties that make the material interesting.

I see the economic rationale as the biggest obstacle to the use of organic and biodegradable materials because the business case is not so clear yet, and you have to compete with very cost-effective and scalable materials such as plastics. You can say a lot of bad things about plastics, but it is very hard to beat. But for the most part, we have what we need to make biodegradable or edible robots, or at least robots with just a few nonedible components. The big gap still lies in the integration of electronics and energy sources, the hardest part to make biodegradable. There are solutions but not ready for prime time yet.

Electronics play a key role in conventionally made robots, but it turns out that several food materials have electrical properties that can be leveraged for assembling edible and even nutritional electronics.[6] All electronic devices are made of five types of components: insulators (do not allow current flow), conductors (allow current flow), semiconductors (materials with both insulating and conductive properties, for example to allow charge carriers to flow only in one direction), sensors (translate environmental stimulation into current), and batteries (energy storage).

Most food materials, such as cereals, bread, oil, gelatin, and dried vegetables, are good insulators and have different mechanical properties that make them suitable for specific uses. For example, rice paper could be used as a substrate for conductive wires and interconnections because it is thin and flexible; sweet potato starch could be used as substrate for resistors because it can be modeled in various shapes; hard gelatin can provide a substrate for a multitude of edible electronic components and shaped as a tablet or a pill; and edible cellulose can be used for larger substrates.

However, few food materials have conductive properties within the value range typically used in conventional electronics. The majority of food materials display none or very low electron conductivity, with the exception of carbonized sugar, carbonized cotton, and carbonized silk, whose conductive power is at best 100,000 times smaller than metallic wires. However, it is possible to use gold and silver, which are edible metals that have excellent conductivity and are commonly used in conventional electronics, for wires and interconnections. In particular, high-purity gold (higher than 23 karats) is a commonly used food additive that can pass through the digestive tract without being absorbed. If released in the environment, gold does not degrade, but it is not toxic either. It can be deposited on edible substrates in very small quantities with a variety of methods to print electrical circuits on diverse edible materials. A larger variety of food materials, such as fruits and vegetables, instead display ionic conductivity where ions, and not electrons, can flow through electrolytic substances, such as salts and acids, dissolved in a liquid, such as water. This type of conductivity can be used for AC current, but cellular walls must be broken by crushing or cooking in order to enable ion flow.

Semiconductors are used in electronics for a variety of purposes, such as light emission and conversion of light to current. Several natural pigments derived from plants and animals, as well as artificial pigments authorized for ingestion, display semiconductive properties, although their use in edible electronics and safety studies is less advanced.[7] Some food materials display piezoelectric properties that convert mechanical stress into electrical current, such as cellulose, which is found in many vegetables such as broccoli and brussels sprouts.

The realization of edible sensors requires the combination of multiple food materials. For example, broccoli powder was blended with gelatin and an edible plasticizer to mold a piezoelectric sensor into a desired shape with desired elasticity; when coated with gold electrodes on the two sides and laid flat on a loudspeaker's cone, the edible sensor could pick up and reliably transmit to an amplifier the sound corresponding to the bowel movements of a person, paving the way for edible microphones for diagnosis of digestive problems.[8]

Although the field of biodegradable and edible electronics is developing rapidly, it has not yet made it into fully edible robots.[9] As Omenetto emphasized, edible batteries are the most challenging component because there are few food materials that can be used for replacing the nonedible chemistry of conventional batteries. Furthermore, since power storage is proportional to the battery mass, inefficient edible batteries could be disproportionately larger than the rest of the robot's body. However, edible batteries will have lower requirements than conventional batteries: their lifetime will be only a few days and will be dictated by the perishability of the other food components of the edible robot, their use will be limited to a few minutes or hours while the robot is being eaten or traverses the gastrointestinal tract, and the current output will necessarily be low in order to prevent electrical shocks when eaten. With these features in mind, researchers turned their attention to supercapacitors, which have lower power storage and current generation but can be recharged faster than a conventional battery. For example, researchers showed an edible supercapacitor capable of storing up to 3200 watts per kilogram made of layers of active charcoal separated by a layer of seaweed with Gatorade serving as electrolyte, with the outer layers made of cheese and packaged within a gelatin envelope.[10] Another promising battery technology is microbial fuel cells made of edible and safe microorganisms that produce charged ions by consuming an edible material. For example, a microbial fuel cell made of the yeast *Saccharomyces cerevisiae* feeding on glucose was capable of producing 0.39 volts for two days, but it included a no-edible proton exchange membrane.[11]

Microbial fuel cells have been shown to produce sufficient energy for driving robotic actuators by intermittent current pulses.[12] To the best of our knowledge, Slugbot at the University of Bristol was the first proof-of-concept robot designed to capture and obtain electrical energy from the fermentation of organic materials, specifically slugs. Slugs were chosen because a slow-moving robot has a better chance of capturing them and because they are considered pests in agriculture and raise less ethical concerns. Slugbot, which was nominated by *Time* magazine as one of the best inventions of 2001, was a four-wheeled robot with a mobile arm equipped with a camera. As it moved through the lettuce field, it scanned the ground with the arm and sucked in slugs detected by a computer vision algorithm. The slugs were deposited in a tank and periodically returned to a fermentation and recharging station located in the field.[13] The same research group went on to develop EcoBots, a series of robots with onboard digestive and power systems. The latest incarnation, Ecobot-III, had a circular stack of forty-eight microbial fuel cells fed with a sludge of dead flies and water. The robot moved on rails between two feeding stations. The energy produced by the microbial cells was sufficient to drive the robot between a water station and a refill station with fly sludge.[14] Recently, they also came up with the design for a paddling robot powered by a single microbial fuel cell feeding on wastewater.[15] The 20-centimeter-long Rowbot could float on water by means of four static legs with polystyrene feet and advanced by rowing with two side paddles. The robot was equipped with a frontal mouth that could open to capture wastewater for the microbial cell and with a fuel exhaust on the back. The microbial fuel cell was separately built and shown to produce more energy than that required by the robot, but the robot and the microbial fuel cell were not integrated and tested.

More recently, researchers have been exploring robots made of edible materials powered by alternative energy sources. For example, the rapid size expansion of heated popcorn was leveraged to fabricate one-time robotic actuators.[16] Popcorn grains encapsulated in tubes with diverse elasticity on the two sides caused the structure to bend when heated above 200 degrees by an embedded filamental heater. A three-fingered popcorn gripper wrapped around and held a ball when heated, but the action was not reversible. Gelatin, an animal-derived protein commonly used in the food industry as a gelling agent and in the pharmaceutical industry for making drug capsules, has attracted much attention by roboticists for its versatility.[17] Gelatin can display electroactive properties by absorbing water with electrical charges and thus bend when positioned in an electric field.[18] It can be mixed with water or other edible substances to obtain the desired consistency and elasticity. For example, in our lab at EPFL we mixed gelatin with glycerol to manufacture an edible pneumatic gripper that displayed mechanical properties and actuation cycles similar to those of conventional soft grippers.[19] When pressurized at 25 kilopascals, its two Toblerone-shaped fingers immediately bent up to 180 degrees and could firmly pick up objects of different shapes and consistencies, such as apples, oranges, and Lego bricks. In another example, MIT researchers fabricated gelatin-glycerol structures with a chamber filled by this reactant powder and a slit that served as an exhaust valve.[20] When water was injected into the cavity with a syringe, the resulting inflation caused a deformation of the gelatin-glycerol structure until the slit opened to let out gas and the structure returned to the original shape. Although the authors showed that a combination of two cylindrical structures with opposite chamber locations produce a rolling motion, the generation of continuous and sustained motion remains an open research challenge at the time of writing this book.

HUNAN PROVINCE, CHINA, 2051

Javier had rarely been hungrier in his life. What a morning it had been, after all—getting up at 4:00 a.m., traveling by bus for three hours from the conference center to the park, getting on and off the jeep to tour the park, and walking under the sun for a few kilometers. He was prepared to devour anything he would find on its plate, but he was not fully prepared for what was on the table.

When the waiter put the plate in front of him, there were a dozen pieces arranged in a circle on the edge of the plate. They had various shapes, some geometric and some irregular. Some looked like bread, others like jelly, and others like cheese, potatoes, and meat. When the server used a dropper to place a single water drop on each piece, they all came alive, started to twirl and move on the plate, and assembled into what looked like a miniature forest landscape. "Our dishes are meant to represent the very ecosystems from which the flavors, and constituting ingredients, come from," the server said.

Next to their table, Song, the restaurant's famous chef whom Javier had seen on the cover of a fine dining magazine a couple of months before, was explaining what was happening on their plates. "In your case, sir, it's the forest," he said to Javier. "For you, madam, it's that unique place where the river meets the sea, and freshwater and saltwater mix," he said as he turned to Corinne, who was witnessing the food in her plate form the shape of a river delta. "Yours is a valley between the mountains," he charmingly said to Aya, "while yours," he said to Katsuo, who had a deep bowl instead of a plate, with food swimming and swirling in it "is of course the open sea." Katsuo seemed less shocked by his plate than the others. After all, eating raw or live fish, shrimps, or other seafood was relatively common in several Asian food cultures, and in some cases—such as with octopus and shrimps—that included eating the animal while it still moved. As for Mei, she had obviously

dined at that restaurant several times in the past and was just enjoying the effect it was having on their guests.

"Robotic food," the chef continued "is the other end of the technological spectrum of what you've witnessed in the park. The same logic that allows Mei and her colleagues to create artificial animals that move and get eaten by real ones, allows me and my team to create food that is active—that moves, changes shape, changes texture and color as it receives physical and chemical signals from the environment—be it the plate, your mouth, your stomach. We take the core components of food such as carbohydrates, starches, proteins, lipids. We use them to build active parts or small sensors that we incorporate in structures that sometimes resemble traditional food and sometimes look like something entirely new. And we program them so that they do something visually interesting when they get on your table—or simply to fully release their flavor when they are in your mouth."

"We are very proud to have Song here in our little restaurant," said Mei. "He was awarded the second Michelin star last month, as you may have heard—the first such honor for a chef working with robotic food."

"It's a big honor of course," said Song, "mostly because it means these techniques are now being accepted into the food culture. At first, they were vehemently opposed, like it happened to Ferran Adria's molecular gastronomy at the end of the last century. But by the early 2000s, his restaurant in Spain was collecting prizes upon prizes, and now that kind of cuisine is commonplace. We like to think we are continuing that sort of innovation in food, by breaking the last barrier: making food active instead of passive."

LAUSANNE, SWITZERLAND; BOSTON, MASSACHUSETTS, USA; AND WOLLONGONG, AUSTRALIA, CURRENT DAY

Food and robots have rubbed shoulders for a while, from harvesting by machines, to automated cleaning and processing, all the way to robotic packaging and autonomous delivery of ready-made dishes by ground and air vehicles. Today, robots are employed in the food industry mainly to accelerate the transition from the field to the plate, improve hygiene, and reduce costs. Food is also 3D-printed by a variety of computer-driven technologies[21] to obtain complex textures and flavors, and recipes can be personalized by apps.[22]

But robotic food, as we envision it, holds the potential of bringing raw food ingredients back to life for new experiences, addressing nutritional and digestive disorders, reducing waste, and promoting sustainable food sources. Robotic food is food that reacts to environmental stimulation to produce a behavior or a sequence of behaviors, such as motion, color, texture, and flavor change, or more complex behaviors such as self-assembly into novel superfood. To this end, we have recently started a long-term collaborative research program that brings together experts in edible electronics, edible actuators, and food production.[23] We expect that edible robots and robotic food will share many technologies and design principles, but the main purpose and function of robotic food will be to provide healthy, appetizing nutrition. Also, while edible robots will mainly operate in open environments, robotic food will make its magic mostly in kitchens, house and restaurant tables, and in our—or animals'—mouths.

We are still far from the scenario described in our fictional story, but there are already promising results. Researchers at the MIT Media Lab developed shape-changing food that folds from thin, flat films into complex 3D patterns while cooking, such as gelatin-cellulose films and flour-based pasta.[24] The principle is similar to that of plant movement described earlier in this chapter: flat pasta is patterned with surface grooves that produce an asymmetric swelling when placed in boiling water, resulting in complex 3D shapes that are visually appealing and better hold sauce.[25] Furthermore, since the dry pasta is flat, it can be packaged and transported more efficiently than conventional 3D-shaped pasta. They also showed humidity-driven folding of flat films covered by a layer of natto cells, a staple of Japanese food, that expand and deform when exposed to vapor produced by a cup of tea or human breath.[26] The team published a draft of organic ingredients that can change shape, color, odor, and flavor when they get in contact with liquids of specific acidity, thus paving the way for more interactive and tasty experiences with robotic food.[27]

Food bites that swim in your soup may soon be available in restaurants. Researchers at the MIT Mechanical Engineering Department and culinary experts at ThinkFoodGroup company took inspiration from semiaquatic insects that propel themselves in water by releasing tiny drops of a fatty liquid. The liquid reduces water tension at the surface, and as a result, the insect is pulled forward to a patch of water with higher surface tension. The engineers obtained a similar effect with high-grade alcohol released from tiny vessels floating on nonalcoholic drinks, and the culinary experts made the vessels edible and tasty.[28] The cocktail boats, just 1.5 centimeters, filled with Bacardi rum (75% alcohol), could swim in a cup for up to two minutes.

The field of robotic food is young and promising, but for the moment, researchers have been experimenting with combinations of edible and nonedible materials. The use of nonedible parts in food is not new: for example, they are often used for decoration and entertainment, such as the candles on a birthday cake. Researchers in Wollongong, Australia, proposed an interactive breakfast "breadboard" (in electronics, a breadboard is a flat base with connectors for prototyping electronic circuits). They used Marmite breakfast spread, a good ionic conductor, to print an electrical circuit on a slice of toasted bread capable of switching on and off an LED by drawing current from an external nonedible battery.[29] One could envision a future where children learn to draw electronic circuits on delicious and nutritious breakfast cakes. Researchers in Tokyo instead turned common beverages, such as coffee and tea, into interactive displays that communicate with the drinker by producing images made of tiny bubbles. Surface bubbles were produced through electrolysis by a battery-powered electronic circuit composed of a matrix of bubble-generating electrodes at the base of the cup.[30] The "bubble pixels," which could be programmed as a display, were connected to an external camera to reproduce the face of the drinker on the liquid surface.[31] In the future, one could envision smart electronics that measure properties of the drink, such as alcohol content and volume, to tell a drinker when it is no longer safe to drive or produce other interactive messages. Other Japanese researchers and artists exhibited the "dying robot," a vegetable body coupled to a nonedible soft actuator lying on a bed of fermented rice bran.[32] While alive, the robot contorted and stirred the fermented rice, thus causing a gradual fermentation and degradation of the vegetable body, making it tastier while slowly dying. In a recent collaboration between EPFL, the Lausanne School of Art, and the Lausanne Hospitality School, we looked at a more cheerful experience with robotic food and modified our gelatin-glycerol recipe previously used for the edible gripper by adding flavors and colors. We then turned the candy-flavored material into a flower-shaped dessert connected to a soda cartridge whose pressure would cause the petals to open so that the person could eat them. As of today, robotic food is a form of experimental cuisine to explore new interactions and culinary experiences. However, it can serve healthy and nutritional purposes as more edible materials with functional properties and more design methods will be available. We also believe that it will serve new functionalities for animals, for example, to make pet food more appetizing while reducing unhealthy ingredients and to deliver medications or vaccines to wildlife animals.

HUNAN PROVINCE, 2051

Around the table, even the most resistant guests had by this point fallen in love with the robotic food. It was beyond delicious, and what happened in the mouth was much more impressive than what had happened before on the plate. They had never experienced such full, satisfactory, and well-defined taste. A single bite corresponded to a sequence of sensations that seemed to come exactly in the right order, each new taste completing the previous one. As Song had explained, the actuation sequence was triggered by chewing and by the chemical environment of the mouth, which caused bubbles of micro-ingredients embedded in the food to burst and release their content in a precise sequence.

The food and the wine had put everyone in a good mood, and the conversation had quickly gone from the park's animals and the restaurant's menu to their philosophical implications.

"Does the distinction even make sense anymore?" Aya asked passionately. "I mean, already this morning we had problems telling what was real from what was not. Some of those prey we saw out there can move around, explore the environment, feed on other organisms. That's what living things do, right?"

"But they can't reproduce, and they can be switched on and off literally with a click of the fingers, which is not something you can do with a live animal," Mei noted.

"Sure," Corinne continued, "but with other technologies, things are way more blurred. Take the colony of robotic . . . I don't know how to call them, tadpoles? The ones that they are using to clean beaches from microplastics in Thailand. They are entirely made of frog cells that are sustained by the same metabolism that sustains real cells in real frogs and communicate with each other more or less like real cells do. And once you kill them, you kill them. There's no way to bring them back. They do seem alive to me; they're just alive in a way that evolution has not come up with."

"They are life as it could be, " quipped Mei. Most of the people at that table knew that quote. It was from Christopher Langton, the founder of research in artificial life, who in the 1980s wrote that A-life, which at the time focused on computer simulations of living agents, could not only help understand life as it was but also life as it could be.[33] "You guys weren't even born at the time," Mei noted with a smile. "I was in high school, and I remember reading the first papers on Xenobots and thinking that they were a turning point. From then on, the relationship between engineering and biology was to change forever. In fact, those papers were a big inspiration for my studies. I might not be doing this job if it was not for them."

BURLINGTON, VERMONT, USA, CURRENT DAY

Since the early 1990s, roboticists have considered the possibility of leveraging the sensory-motor capabilities of living organisms to build robots that take the best of the biological and artificial worlds. For example, scientists at the University of Tokyo used the antennas of a silk moth as sensitive odor detectors for a wheeled mobile robot.[34]. The electrical signals from the biological sensors were translated into digital signals for an artificial neural network that made the robot follow odor plumes as the insect did. A few years later, the same team described the world-first insect borg, a cockroach implanted with an electronic backpack that could be steered by remote control.[35] The backpack contained a wireless receiver, a lithium battery, a microcontroller, and two electrodes for electrically stimulating the nerve bundles of the left and right antennas. The software leveraged the insect's innate escape behavior of moving in the direction opposite to the stimulated antennas. The researchers could also make the insect follow a black line on the ground by adding a photosensor on each side of the body. If a sensor detected the black line, the backpack would stimulate the antenna on the opposite side to steer the insect back on track. However, the insect borg was not very reliable because the electrical stimulation produced by the backpack did not match the amplitude, frequency, and pattern normally received by the biological antennas. Over the past twenty years, these limitations have been gradually removed, and recently a team of researchers at Nanyang Technological University in Singapore revealed a 0.5 gram beetle implanted with a 0.45 gram backpack capable of wirelessly steering the insect for five days, walking up to 1.15 kilometers at an average speed of 4 centimeters per second.[36]. The miniaturization of the electronic backpack and a better understanding of the electrical stimulation allowed the scientists to better grade the steering response and even make the insect move backward with an 85 percent overall reliability. While these biohybrid robots could be helpful in life-saving situations, such as in search-and-rescue through rubble, they could also raise ethical issues.

A different approach consists in harnessing intrinsic properties of biological cells. Here the goal is to produce functional organs, such as muscles and sensors, made of cellular aggregates that can be customized and integrated in robots. For example, researchers at Harvard University developed thin muscular films composed of a layer of cardiac cells, which display spontaneous rhythmic contractions, deposited on an elastic patch covered by extracellular proteins.[37] In a few hours, the layer of cardiac cells interconnected and synchronized their contractions to produce coordinated motion of the structure. The shape of the muscular thin film could be tailored to produce ground locomotion or swimming in water. Another group at Harvard later revealed a swimming robot powered by four layers of cardiac cell tissues that mimics the way ray fish swim by combining two contractile movements: one along the front-to-back axis that results in a waving motion and one in the radial direction expanding outward from the center that can also be used for steering sideways.[38] The cardiac cells were engineered to react to blue light by producing ion currents that would contract the cell and propagate to neighboring cells. The size of the resulting robot was just 2 millimeters (one-tenth the size of the real fish) and was capable of swimming at an average speed of 1.5 millimeters per second over a distance of 250 millimeters. Furthermore, the robot could be steered in a pool by making it follow a blue light projected in the water.

The rhythmic motion of cardiac cells is not suitable for all types of locomotion, however. Researchers at the University of Illinois at Urbana-Champaign and MIT described a method to 3D-print biohybrid robots driven by skeletal muscle cells that contract only when electrically stimulated, thus enabling precise control of movement only when needed.[39] The robot was made of two parts: a 3D-printed bridge structure with two rigid pillars connected by a flexible beam and a gel matrix connecting the bases of the two pillars that was imbued with a solution of skeletal muscle cells and fibrin collagen proteins. The cell matrix was then infused with a growth hormone that caused the cells to interconnect through the fibrin proteins and form a muscle fiber bundle, a process similar to that observed in the formation of skeletal muscle fibers. The resulting muscle strip contracted when electrically stimulated, causing an arching of the bridge. By making one of the two pillars more elastic, the researchers showed that a 5 millimeter robot would produce crawling locomotion at a speed of 0.15 millimeters per second. Interestingly, in nature, skeletal muscle cells connect also to other types of biological cells, such as nerve cells, and could be used in the future to create more complex biohybrid robots with a biological nervous system.

However, compared to biological organisms, those biohybrid robots still have limitations of traditional robots: they cannot adapt, they do not have a metabolic system to feed and produce energy, and they cannot heal after damage. Enter the xenobots, living robots made of assemblies of stem cells and muscle cells from the embryos of the frog *Xenopus laevis*. A collaboration between roboticist Josh Bongard at the University of Vermont and biologist Michal Levin at Tufts University, xenobots are multicellular robots whose morphology is designed by computer algorithms that mimic evolutionary processes to generate robots with desired functionalities.[40] The robots are entirely made of pluripotent living cells that move, produce their own energy, communicate, and adapt to the surrounding environment.

A computer scientist by training and a coauthor of a seminal book on bioinspired robotics, Bongard describes the work that led to xenobots as an act of "intellectual bravery," in particular from his biologist colleagues, who were willing to ask questions few biologists had asked before.[41] "Cells communicate via electric signals, even if they are not neuronal cells," he says. "What we're learning is that by rearranging biological tissues, you rearrange bioelectric communication on patches of tissues and that this morphological rearrangement can be seen as a form of programming. It is very different from the neural approach where you train an animal, or a neural network, and influence synaptic plasticity. This is still high-level programming but a very different one."

The researchers started with a computer model of a three-dimensional assembly of cells where the number of cells, positions in body space, and individual cell properties (for example, support cell, motor cell) were encoded as a string of values, which represented the artificial genome of the simulated robotic organism. From there, the computer created an initial population of one hundred diverse genomes corresponding to different simulated organisms, each displaying a different behavior resulting from the specific layout and cell type. The researchers defined a performance measure, also known as fitness function, to let the computer assess the behavior of the robots placed in a simulated environment and select the best ones for reproduction. Reproduction consisted in making slightly mutated copies of the genomes in numbers proportional to their fitness to create the next generation of robots, which were assessed; only the best ones reproduced and mutated. After several generations of selection, reproduction, and mutation, artificial evolution came up with robot morphologies that obtained high fitness scores. The researchers repeated evolutionary experiments with four different fitness functions selecting for locomotion, object manipulation, object transport, and collective behavior, respectively. They then took the genome of the best evolved individual for each selected lineage and used it as a blueprint to assemble the xenobot out of stem cells and cardiac cells from frog embryos.

Finally, the machine-designed organisms were liberated in small dishes filled with water and particles where they continued to live for up to ten days while displaying the behavior for which they had evolved. When the behavior of the living organisms was different from that observed in simulation due to discrepancies between the model and the reality, the researchers modified the computer model to incorporate the observed constraints and evolved again the xenobot morphologies, resulting in better performance of the real robots. The automated design method used by xenobot researchers, also known as evolutionary robotics, has been frequently used to explore and find solutions in the complex neuro-morpho-sensory-motor space of robotic systems.[42] In several cases, evolved robots were 3D-printed and infused with the evolved control system.[43] However, xenobots were the first living organisms designed by machine evolution. For Bongard,

> One of the most exciting things about xenobots is that they provide a way to investigate life as it could be—which is a quote from Christopher Langton, the founder of the field of artificial life. Most of the work on artificial life for thirty or thirty-five years has been restricted to simple computer simulations. Now that we can manipulate living tissue much better, we can ask those questions using living things. Can we make biological systems that are very different from what evolved? How far can we go? That's where we're going to learn the most about biology and evolution. There's this idea in evolutionary robots to run evolutionary algorithms on mechanical parts, sensors and motors and batteries. The irony to me is that those things are dumb, are modular but not very good at pairing with other things. We've had some success with that strategy, but the fact that we tried evolutionary algorithms on frog tissue for just one year and we got to this point, that changes the rules of the game.

In more recent work, the authors explored the self-organization capabilities of xenobots.[44] Instead of assembling them from evolved genetic instructions, they excised a patch of undifferentiated stem cells from the frog embryo, placed them to heal for an hour in a water solution, and cultured them for four days until they formed a circular shape of approximately three thousand cells that started to move by a combination of rotatory and forward motion. Movement was generated by cilia grown on the peripheral cells. When the xenobot body was surgically lacerated, it could self-heal and regenerate the original shape in only five minutes and resume its motion. Furthermore, when many xenobots were positioned in a dish with inert particles, the robots would capture the dirt and accumulate it in small heaps, a collective phenomenon also observed in ants that result from the interaction between the moving agents and the particles.

"The very first application will be a typical application of robotics," says Bongard. "This is basically frog tissue, so we envision xenobots that inspect underwater machinery such as submarine engines, filters, and wave farms. They could flow into underwater machinery and report back on grime, damage, wear and tear, blockage. That would be relatively straightforward. You see hints of that in the article, where we used red and green fluorescence by which xenobots could in principle see, remember, go back, and tell."

Another application will be environmental remediation. "These things seem to be able to grab onto very small particulate matter. They could find contaminants in soil and bring it to the surface or remove beans of microplastics from water." Of course, at that point, Bongard acknowledges, we would need very detailed regulation, because we would be releasing new biological materials into the ecosystem.

Still, Bongard points out, it's important to keep in mind that biohybrids do not reproduce. This makes scaling the technology up difficult, but makes it more controllable than, for example, genetically modified organisms such as plastic-eating bacteria. "In xenobots, we could consider adding reproductive capacity but in a more controlled manner. We could design and channel their reproductive behavior so that they reproduce only in very controlled conditions, with a very simple kill switch. Of course, once you do that, there is the prospect of evolution, and evolution is great at working around kill switches. It would not be completely safe, but I would argue safer than genetic engineering." A few months after this statement, Bongard and Levin's team reported the first self-reproduction of xenobots.[45] The researchers evolved in simulation xenobot cell assemblies that were capable of engulfing other cells, akin to the game of *Pac-Man*, and shape them into a copy of themselves. These copies would then be released from the parent's body and further self-reproduce with the same method. The specifications evolved in computer simulations were then applied to the biological xenobots, which displayed the same behavior, leading to a chain reaction of self-reproduction events.

HUNAN PROVINCE, CHINA, 2051

The conversation had been going on for almost an hour, interrupted at times by the increasingly surprising dishes that composed the tasting menu. The lunch ended with an ice-cream-based dessert that literally formed from its room-temperature ingredients in front of their eyes. Seeing all that active stuff on the table somehow added fuel to the discussion about the meaning, and the ethics, of creating machines out of biological material and releasing them into the environment.

In the end, the group thanked Mei for the experience and for lunch, shook hands, and headed outside. Indeed, they had to get back on the bus if they wanted to catch some of the late afternoon talks in the conference that had brought them to Hunan. The conference, the world's largest on bioinspired technologies, had been launched ten years earlier as a way to encourage interaction between engineers and biologists when the distinction still made sense and communities were still very much separated in academia. Now, most participants to the conference—the four of them included—would have had problems labeling themselves one way or the other. They were just scientists who studied and built intelligent systems, using any tool they could find.

10 HOW TO COMPETE WITH ROBOTS

We have made it to this point in this book without ever mentioning the origin of the word *robot*. As some of our readers may already know, it comes from a work of fiction, the play *R.U.R.—Rossum's Universal Robots* by the Czech writer Karel Čapek. Written in 1920 and first performed in Prague the following year, the play depicts an imaginary future set in the year 2000, where artificial creatures called—guess what—"robots" are cheap and ubiquitous and are used to replace human workers in factories around the world. The play's robots are not mechanical things: they are synthetic human beings, mass-produced thanks to a mysterious chemical that allows humans to build living tissue and quickly grow complete organisms. Čapek derived the word *robot* from the Slavic term *robota*, used to indicate the peasants who were forced to do hard and compulsory work under the feudal system. The play does not end well for humans: robots rebel and take over the world, killing almost all men and women in the process. But for a while before the robot revolution, humans seem mostly happy to be freed from work. The play mentions episodes, before the time when the story takes place, of American workers who resisted being replaced and smashed the robots—but suggests that the rest of the population sided with the robots and even armed them against the rebels.

In other words, from the very outset and from the very choice of the word, the concept of "robot" is somehow built around the idea of replacing human work. And even Čapek, whose take on the consequences of automation was in the end fairly pessimistic, seemed to think that humans would be overall happy to forgo work, at least in the short term.

A century later, Čapek's concerns about robots and artificial intelligence taking over and turning humans into slaves are still staples of science fiction, but are dismissed as nonsense by most scientists. However, concerns about the immediate social consequences of automation abound. They are intensely debated by engineers, economists, and sociologists—most of whom agree that massive replacement of human workers by machines would not go down so smoothly.

Several best-selling books have depicted the next few decades as the time of "the rise of the robots" or of the human "race against the machine," while studies predicting how many jobs could be lost to automation are churned out on a regular basis.

The impact of technological change on human jobs is nothing new. Historical examples already abounded before Čapek wrote his play and continued throughout the twentieth century. The term *luddite*, which has become synonymous with hostility against new technologies, comes from a movement of workers opposing (and sabotaging) the introduction of mechanical technologies in British textile factories in the early nineteenth century. In the early 1900s, the mechanization of agriculture in rich countries caused a massive displacement of workers toward other sectors: the share of the US workforce working in farms and fields declined from 50 percent in 1870 to just 4 percent in 1980, largely as a consequence of mechanization.

Throughout history, technology has both created and destroyed jobs. The early years of the telephone created the job of switchboard operator, which was erased after a few decades by automatic switching. Typewriters demanded professional typists—until they were replaced by personal computers. Overall, though, the net balance has been positive. Technology—in particular, the two waves of spectacular innovation associated with the first and the second industrial revolutions—has accelerated economic growth up to a pace that has no precedent in human history. Machines did take some jobs that were previously done by people—Luddites had a point after all. But they created many more new jobs. Automatized textile factories could produce much more at a lower price. By selling more items, they grew bigger and ultimately employed more people who attended to power looms and take care of other stages of textile productions. Electrification and the internal combustion engines—the same technologies that threw millions of people out of jobs in agriculture in the early twentieth century—at the same time created new jobs in the industrial sector. Unemployed American farmers moved to cities and became factory workers. None of this happened smoothly. In fact, these shifts were often brutal, and the changes were not always for the best. At various times in the twentieth century, periods of recession left millions unemployed and without an income. But overall, technological change during the first two industrial revolutions did not directly cause long-term mass unemployment.

But what about the third industrial revolution—the one brought about by computer and information technology—and the fourth one that we are depicting in this book, based on advanced artificial intelligence and bioinspired technologies? According to several influential authors, the impact of technology on jobs will be very different this time.

One of the first authors to brush up the idea of a Čapekian world where humans no longer need to work was Jeremy Rifkin, in his 1995 book, *The End of Work*.[1] An activist and social critic known for his visionary and often extreme takes on what the future will look like, Rifkin pointed out that in the first two industrial revolutions, it was physical work that was being replaced. Humans could still use education—for themselves or for their children—to access jobs that no machine could do. The second half of the twentieth century did indeed witness a massive shift of workforce from the industrial sector toward the service one. But with the third industrial revolution, machines started to take over some cognitive tasks too and threaten white-collar jobs, not only blue-collar ones. Rifkin saw a new era approaching, "one in which fewer and fewer workers will be needed to produce the goods and services for the global population." He went as far as recommending policies to cope with this sea change—from imposing a shorter workweek in order to give jobs to more people, to states massively subsidizing charity and social work, simultaneously assisting those on the losing end of technological change and keeping people busy. Rifkin wrote his book in the mid-1990s, before the rise of the Internet and well before the recent advancements in artificial intelligence, made possible by last-generation neural networks and deep learning. He does deserve credit for anticipating a debate that would grow in significance a couple of decades later.

The idea that machines would disrupt—if not entirely end—human work became mainstream in the 2010s. One of the most influential books in that period was *Race against the Machine*, published in 2011 by two MIT professors, Erik Brynjolfsson and Andrew McAfee.[2] In their view, the main engine driving the transformation is Moore's law, which predicts—in its most popular form—that the number of transistors that can be fitted on an integrated circuit doubles every eighteen months. This means that computing power grows at an exponential pace, which leads the two authors to predict that no matter how spectacular the growth in computer performance has been in the past couple of decades, we have seen nothing yet. To explain why, they use the old story of the emperor who wants to reward the inventor of the game of chess and asks him what he wants. The inventor replies by asking for rice, in a quantity to be determined as follows: one grain on the first square of the chessboard, two grains on the second, four on the third, and so on, doubling the grains at each square until the chessboard is complete. The emperor accepts, thinking he had the better part of the deal—until he discovers the perils of exponential growth (this anecdote became popular again during the COVID-19 pandemic, to explain the danger of letting infection cases grow exponentially). Computers and artificial intelligence, Brynjolfsson and McAfee argue, have entered the second half of the chessboard, where every improvement is dramatically more significant than all of those that came before. As a result, the two authors expect machines (in fact, computers and AI) to quickly become capable of taking up more and more tasks in the service sector—performing crucial bits of customer service, consultancy, and other jobs where humans are still considered necessary.

The two professors are slightly less pessimistic than Rifkin. They do not believe that the end of work is near, though they do believe that the unemployment rates seen in the United States in the 2000s could indeed be attributed to automation. They believe AI will ultimately boost productivity and growth, leading to new industries and ultimately to more opportunities for humans. But they suggest that this will take time and that one or two generations of low-skilled workers will find themselves caught in the middle, without having time to adapt. Humanity will ultimately win the "race with the machine," but some workers are already losing it. Like Rifkin, they suggest correctives: promoting organizational innovation in companies so that entrepreneurs find creative (and profitable) ways to let medium- and low-skilled workers work together with advanced technologies, so that the result is bigger than the sum of the parts. And education, education, education: online courses and other novel educational technologies, they believe, offer a chance to train millions of students of any age faster than in the past, and in particular to teach them how to stay ahead of innovation, effectively training designers, rather than users, of technologies.

But in *Rise of the Robots*, Martin Ford thinks that this time is different.[3] He notes that between the end of World War II and 1970, productivity and income kept rising in sync in most Western economies. During that period, technology made production more efficient and earned larger profits for entrepreneurs and better salaries for workers. But after 1970 something broke. Ford describes "deadly trends" that have plagued the economies of the United States and other developed countries since then. Wages have remained stagnant. Labor's share of national income (in other words, the part that turns into salaries) has "plummeted," while the share going to corporate profits has "skyrocketed." Fewer jobs have been created, and long-term unemployment has become more common. Income inequality has accelerated. Ford quotes economic research according to which, between 2009 and 2012, 95 percent of income gains in the United States went to only 1 percent of the population, as opposed to 50 percent between 1993 and 2010. The job market has become increasingly divided between "low-wage service jobs and high-skill, professional jobs that are generally unattainable for most of the workforce." Part-time jobs, once an exception, have become more common—in most cases not because workers choose them but because they cannot find an equivalent full-time post.

These trends can be attributed to many things, Ford acknowledges, from the oil crisis of the 1970s to globalization of finance and logistics of current days. But overall, argues Ford, there is clear evidence pointing to information technology as a "disruptive economic force" that is largely contributing to these trends. Ford is more interested in physical robots than are Rifkin or the Erik Brynjolfsson and Andrew McAfee duo, who mostly analyze the impact of software and AI algorithms. Ford, on the contrary, foresees a wave of rapid advancement in industrial and service robots. In particular, he mentions the widespread adoption of ROS (Robot Operating System), a free and open source operating system that has become a de facto standard in the robotics community. In desktop and mobile computing, Ford notes, the convergence toward one or two standard operating systems has greatly accelerated the development of applications, and he bets that the same will happen to robots because of ROS. He also sees a major agent of change in cloud robotics, the technology that lets robots connect to the Internet to access computation and data. That will require less onboard memory and computing power, making robots lighter and less expensive. It will also allow them to be continuously updated and to share data that other robots collect from the environment and what they learn from their work.

As a result, he foresees a steep rise in the use of robots in the manufacturing, service, and agricultural sectors that could take away millions of manual jobs in food production, retail, logistics, and textile manufacturing. In the meantime, AI and computers will keep destroying white-collar, cognitive jobs at an increasing rate, at some point becoming able to also perform some high-level, quasi-managerial tasks that are now considered completely safe. The future Ford has depicted does not look pretty for human workers, and he does not believe that retraining workers is the silver bullet. "The conventional wisdom is that, by investing in still more education and training, we are going to somehow cram everyone into that shrinking region at the very top," he writes. "I think that assuming this is possible is analogous to believing that, in the wake of the mechanization of agriculture, the majority of displaced farm workers would be able to find jobs driving tractors. The numbers simply don't work." Ford ends his book by arguing in favor of a universal basic income to cope with the inevitable mass unemployment, along with a number of incentives to keep humans busy doing something useful for society, even though for many, that will not be "work" in the traditional sense.

A partial confirmation to Ford's suggestion comes from a study published in early 2021 by a group of researchers at Princeton University and Boston University,[4] Using detailed data on job descriptions and vacancies from forty thousand firms, they assessed how many of the tasks and jobs those firms require can already be done by AI. They then tried to understand how this relates with the number and type of people those firms look for and hire. Firms where a lot of tasks can be done by algorithms tend to hire more people with an AI background, which is not a surprise. But they also tend to hire fewer people overall, contradicting the expectation that increased productivity brought about by the use of AI should translate into more business and more jobs. The authors found this trend at the level of the individual firm but not at the aggregate level. Even between 2014 and 2018, the period where they saw the bigger shift in the behavior of individual firms, there was no detectable overall reduction of labor demand in the United States. The authors interpreted this as a sign that the effect of AI is still too much confined to specific sectors and specific types of firm to be detectable. Still, the trend is clear: if machines do conquer more jobs and more sectors, the effect will also become visible.

Best-selling books aside, predictions about the future impact of automation can also be found in several studies compiled by scholars and research firms. Try yourself, and type "robots and jobs" in a search engine. As we are writing this chapter, the first result is a Bloomberg news story saying that over 800 million workers around the world are threatened by advancements in automated technology.[5] Scroll down, and we find another one stating that 51 million workers in Europe alone risk their jobs within ten years due to competition from robots. The source is a report by the consulting firm McKinsey on the combined effects of automation and COVID-19 on the European labor market.[6] Scroll further down, and we find a somewhat optimistic forecast that by 2037, automation and robotics will cancel 7 million jobs but will also create 7.2 million new ones, for a total gain of 200,000 jobs.[7] That sounds much better, but only if your job is in the right column. If, for example, you work in manufacturing, your job may be among the lost ones. Indeed, according to yet another Oxford Economics research from

2019, which may show up in your Web search, up to 20 million manufacturing jobs will disappear by 2030.[8]

When it comes to predicting the future impact of robotics and artificial intelligence on the job market, choices abound. If this month's prediction seems too gloomy, all you have to do is wait for a better one next month. In 2018, the *MIT Technology Review* compiled a table (alas, already outdated) with all the studies published on the subject of automation and jobs.[9] The results ranged from catastrophic (2 billion places destroyed worldwide by 2030) to optimistic (the International Federation of Robotics expected robots to *create* 3.5 million jobs by 2021). In between, one can find all possible intermediate degrees of disruption.

But where do those numbers come from, and how are they calculated? The first study that tried to methodically assess the number of jobs that could be lost to automation was performed by Carl Frey and Michael Osborne from the University of Oxford.[10] It initially circulated in 2013 as an internal document of the university and was later revised and published in 2017 in the journal *Technological Forecasting and Change*. The most often cited result from this study is that "47 percent of jobs in the United States are at high risk of automation." In particular, 33 percent of American workers therefore have jobs with a low risk of being replaced by a computer, 19 percent are at medium risk, while as many as 47 percent are at high risk of being replaced by an algorithm within a decade or two. According to this study, the administrative, sales, and services sectors are more at risk; health, finance, maintenance, and installation of devices are safer.

Since several following studies on this topic rely on a similar methodology, it's worth understanding how Frey and Osborne got to their numbers. In a first step, they organized a workshop at the University of Oxford with the participation of a group of "machine learning experts." Frey and Osborne gave to the workshop participants a list of seventy jobs accompanied by an extended description extracted from O*net, an American database that encodes and standardizes occupational information for statistical purposes. For each of the seventy jobs, Frey and Osborne asked the experts: "Can the tasks required by this job be specified clearly enough to be performed by a computer?" In short, can you transform that job into an algorithm? The seventy jobs had been chosen in such a way that the answer was easy: surgeon, event organizer, stylist (all difficult to automate); dishwasher, telemarketing operator, court clerk (all easily automated). The authors also asked the experts about the bottlenecks that artificial intelligence and robotics have yet to overcome, and the participants listed three: perception and manipulation, creative intelligence, and social intelligence.

In a second step, the authors looked at the list of abilities required by each of those seventy jobs, as detailed in the O*net database and picked nine that in their opinion best describe the three technological bottlenecks: persuasion, negotiation, social sensitivity, originality, ability to deal with others, knowledge of fine arts, manual dexterity, and ability to work in a cramped environment. Then Frey and Osborne used a machine learning algorithm, of all things, to do the final piece of work for them. They used the seventy jobs analyzed in the workshop as a training set to extract correlations between the importance of those nine abilities in each job, as assigned in the O*net database, and the probability of automation of a job assigned by the workshop participants. Once the algorithm learned the correlations, they used it to compute the automation probability of all the remaining jobs in the O*net database. The resulting list of automation probabilities reveals both expected and unexpected results. For example, at the bottom, with the highest risk of automation, you find jobs as a watch repairer, insurance agent, tailor, and telemarketing operator, all with an automation probability of 0.99. At the top, with the lowest risk, recreational therapists, mechanical machinery installers and maintainers, and director of emergency procedures score automation probabilities below 0.003.

Over the following years, other studies came out—and made headlines—that were more or less an extension and refinement of Frey and Osborne's method. A 2016 study by the Organization for Economic Cooperation and Development used a similar methodology to obtain less alarming results: just 9 percent of the jobs in the organization's member countries were described as automatable, although with important regional differences.[11] For example, 6 percent of jobs in South Korea resulted as automatable contrasted with 12 percent in Austria. A 2017 study by McKinsey relied on the O*net database to break down existing jobs into a list of two thousand activities.[12]

The firm's analysts then defined a list of eighteen "requirements" that each of those activities had to a different degree. This method was similar to Frey and Osborne's characterization of jobs into nine abilities, but the results were more complex and nuanced than those of the other two studies. According to McKinsey, almost half of the activities that make up today's jobs worldwide are potentially automatable. Less than 5 percent are fully automatable, and about 60 percent are composed of at least 30 percent of automatable activities. Finally, the authors observed that the automatable activities concern 1.2 billion workers worldwide and $14.6 million billion in wages.

All of those studies have given an important contribution to the debate surrounding automation. But as roboticists and keen observers of this field of research, we were not totally satisfied with them. First, these calculations contain a lot of subjectivity, ultimately relying on a few experts' assessment of the key requirements that would make a job automatable and of how advanced the replacement technology is. Second, those studies are mostly concerned with algorithms and software versions of artificial intelligence rather than with intelligent robots that move, manipulate, and build things. Third, those studies do not provide indications or advice on the potential costs and benefits of retraining in order to transition to better future-proof jobs.

Working with a team of economists at the University of Lausanne, we made our own attempt at assessing the likely impact of the current wave of automation on jobs.[13] We took into account robotics' physical abilities rather than only software AI; developed a method for objectively assessing robotic capabilities and automation risks; and developed an algorithm that suggests career transitions to more resilient jobs with minimal retraining efforts.

The first thing we needed was a way to compare human and robotic abilities more systematically. On the human side, the O*net database does a great job of classifying approximately a thousand occupations. It breaks down each occupation into dozens of different skills, abilities, and knowledge types with scores that describe how important those skills, abilities, and knowledge are for that job and how good the worker must be at each of them. O*net has been around since the early 2000s and is regularly updated through large-scale surveys with workers and human resources managers.

Engineers would love to have something like that for robots. The resource that comes closest to it is the list of robotic abilities in the European H2020 Robotic Multi-Annual Roadmap (MAR) released by SPARC, a public-private partnership between the European Commission and the European robotics industry.[14] It includes a long list of things that a robot may be asked to do—grouped into ability families such as "perception," "manipulation," and "motion." We went through the two lists and established a conversion system to match O*net human abilities and MAR robotic abilities. The MAR list does not tell us how good robots currently are at each of those abilities, so we went through research papers, patents, and robotic products and made our own assessment using a well-known scale for measuring the level of technology development, also known as the technology readiness level (TRL). The TRL scale goes from 1 (very preliminary ideas only explored in basic research) to 9 (tried and tested technologies ready for deployment). After assigning TRL values to the MAR robotic abilities, we had everything in place to calculate how likely it is that each existing job can be performed by a robot. It is a function of what abilities and skills it requires, how well they are matched by robotic abilities, and how developed those robotic abilities currently are.

We came up with a ranking of almost one thousand jobs, with "Physicists" being the ones who report the lowest risk of being replaced by a machine and "Slaughterers and Meat Packers," who face the highest risk. Although automation risk varies greatly both inside and across job families, jobs in food processing, building and maintenance, construction, and extraction appear to be the ones with the highest risk. This result is consistent with those from Frey and Osborne, although their calculations, reflecting a focus on algorithms and AI, also showed a larger share of jobs at risk in the service and sales sectors.

What we were most interested in, though, was to measure the impact of retraining to shield workers from the risk of becoming obsolete. We thus devised a simple algorithm to find, for any given job, alternative jobs that have a significantly lower automation risk and are reasonably close to the original job in terms of the abilities and knowledge they require, thus keeping the retraining effort minimal and making the career transition feasible

To test how our method would perform in the real world, we used employment data from the 2018 US economy. We divided all occupations in three groups according to their automation risk: high, medium, and low risk. For each occupation, we simulated a move to the occupation suggested by our method, and then calculated for each risk group the average change in automation risk and the average retraining effort to catch up with the required abilities and knowledge. Finally, we weighed each occupation by its percentage of the total US workforce using data from the US Bureau of Labor Statistics.[15]

We saw three encouraging results. First, workers who follow the career recommendations of our method can substantially reduce their automation risk. Second, risk mitigation is possible for nearly all types of occupations. Third, workers in the high-risk group have to undergo a comparatively lower retraining effort in order to move to a safer job.[16]

This method could be used by governments to measure their citizens' risk of losing jobs to intelligent robots and to target educational efforts accordingly. But all of us could one day use it to assess our own automation risk and identify the easiest route to reposition ourselves on the job market. Although this study is only an initial step toward a deeper and systematic study of how robotic technology will affect the job market, it shows that the "race against the machine" can be won. At the very least, it can be fought with realistic retraining efforts.

11 INVENTING AN INDUSTRY

In the first pages of this book, we made it clear that we were not playing in the same league as writers such as Isaac Asimov or Philip K. Dick. First, because we couldn't. We are nowhere as good at creating imaginary future worlds. Second, because we didn't want to. We wanted to imagine a future that is firmly grounded in what we know robots can—at least theoretically—do, which means that we have to stick to the laws of physics and biology, as well as to the core principles of engineering. For science fiction such as Asimov's or Dick's, on the contrary, it is almost a prerequisite to stretch the laws of physics and biology—or at the very least, to imagine new principles that have not been discovered yet and make possible tomorrow what is impossible today.

It is interesting, though, that both Asimov and Dick were a bit less imaginative when it came to describing how future robots will be manufactured and marketed and were less ready to stretch the laws of business than those of physics or biology. One of the most often-recurring characters in Asimov's thirty-seven short stories and six novels on robotics is not a person but a company called U.S. Robots and Mechanical Men.[1] It is the ultimate multinational company—at some point multiplanetary when some of its activities are delocalized into space. It controls most of the market for conscious, "positronic" robots that populate Asimov's fictional universe. In one story at least, Asimov alludes to a competitor, Consolidated Robotics, but it seems to be more or less what Apple was to Microsoft in the 1980s: not a big concern.

As for Dick's Nexus 6 androids, the superrealistic human replicants of the story "Do Androids Dream of Electric Sheep" are produced by a company called the Rosen Association.[2] It is based in Seattle (interestingly, Dick wrote the story in 1968, some years before the city on the Pacific Coast became a technology hub thanks to Microsoft and then Amazon), but outsources production to a Mars colony. In the first *Blade Runner* movie, adapted from Dick's story, the Rosen Association becomes the Tyrell Corporation and is relocated to Los Angeles.

U.S. Robots and the Rosen Association are both big corporations, funded and run by powerful business magnates, with a quasi-monopoly or full monopoly of their market. The future of the robotics industry, as imagined by Asimov and Dick, has a lot in common with the industrial capitalism of the twentieth century when they were writing. If you take out the dystopian element—well, most of it—those fictional companies resemble the Big Three that at that time dominated the US automobile industry (GM, Ford, and Chrysler), the Japanese *zaibatsus* (powerful conglomerates such as Mitsubishi and Mitsui), or the IBM of the 1970s and 1980s when it was practically synonymous with the computer industry.

It is entirely possible that the robotics industry of the future will follow such a trajectory, ending up with a few conglomerates dominating the market and controlling every aspect of their robotic products, from hardware to software. But new technologies do often come with entirely new market structures and business models, especially if they appear at a time when the world market is itself in turmoil, with new superpowers on the rise (China), economic growth in new regions (Southeast Asia and Africa), and new ways of doing business, such as the sharing and gig economies that are replacing purchase and long-term ownership.

The fact is that we don't really know what the robotics business will look like in a few decades. What we can do, then, is to look at where the industry is now, identify the crossroads that will decide its future trajectory, and explore the alternative paths. In this chapter, we do this with the help of experts who study the robotics industry, make business with robots, and help others grow robotics businesses.

OLD AND NEW GUARDS

Let's start from the easiest part: what the robotics business looks like now. Analysts tend to divide it into two worlds of industrial robotics and service robotics. Every year, the International Federation of Robotics publishes two separate reports for the industrial and service sectors, one of the best sources of statistics and information about their markets. These two robotics worlds have very different numbers, players, and business models, and at the risk of oversimplifying, we call them the Old Guard and the New Guard.

The Old Guard consists of industrial robots used to make things or used in factories where things are made. They include the articulated robotic arms that work on assembly lines of car, electronics, and chemical industries where they lift, position, weld, polish, cut, or assemble parts at superhuman precision, speed, and endurance. They also include moving robots that shift components and finished products across factory floors or warehouses.

Industrial robots are the most mature branch of robotics and have developed for more than half a century with a clear business case and deep-pocketed customers that can make long-term investments. As of 2020, more than 3 million robots were in operation in factories around the world.[3] The leading buyers of industrial robots were the automotive industry (32.2 percent), the electrical and electronics sector (25.3 percent), the metal industry (10.3 percent), chemicals and plastics (6.3 percent), and food and beverage (3 percent). Industrial robots generate between $13 and $14 billion in revenues per year. Those are big numbers, but nowhere near the market for computers, worth over $300 billion per year, or the smartphone market, which surpasses $700 billion per year and keeps growing. Let's not even mention the car market, worth over $20 trillion per year. Industrial robotics is, in other words, a flourishing but niche business.

But the picture is not consistently rosy. Even before the COVID-19 pandemic started to upend the world economy in early 2020, the sector had been entering a downturn in 2019 after six years of continuous growth through most of the 2010s. It was a reminder of how closely the industrial robotics business is tied to the fate of its main customers. Bad years for the car industry typically result in bad years for those who provide their robots, and the end of the 2010s were not great times for the car industry, with global demand curbed by a slowing Chinese economy and the rise of shared mobility in many areas of the world.

Today, the structure of the business in industrial robots is well defined. Geographically, Asia is the biggest market for industrial robots, followed by Europe and the Americas. More than 70 percent of all industrial robots go to five countries: China, Japan, the United States, the Republic of Korea, and Germany. China in particular has been the world's largest buyer of industrial robots since 2013. As for who produces industrial robots, there is no Universal Robots or Tyrell corporation, but four companies have historically had such a strong grip on the business to be nicknamed the Big Four: Fanuc and Yaskawa in Japan, KUKA in Germany, and ABB in Sweden and Switzerland. In 2016, Kuka was acquired by the Chinese Midea group, signaling the beginning of a power shift. Other important names in the business are Mitsubishi (a giant Japanese conglomerate active in many areas, from consumer electronics to cars), the German company Bosch, and Epson, another Japanese brand best known for consumer electronics but also active in industrial automation.

The fact that industrial robots represent the most consolidated part of the business does not mean that they are not touched by the technology trends that we have described in this book. In fact, a tidal change is underway in the automotive industry in particular, where the assembly line—for decades, the working environment of industrial robots—is being gradually replaced by a looser organization of the production space, leaving room for both stationary and mobile robots and collaborative robots (cobots). These new industrial robots make it easier to personalize car models because the introduction of a new variant requires mostly reprogramming the robot instead of dismantling and redesigning the whole production line. Collaborative robots that can automatically reduce force or stop when a human gets in the way are now starting to work side by side with humans. Robots are becoming easier to program and integrate in a production system, although there are still several competing standards that do not talk to each other, as was once the case in the computer industry. Cloud services allow robot manufacturers to analyze operations and offer remote maintenance, while machine learning can find patterns in cloud-based data from multiple robots performing the same task, to improve their performance.

The New Guard of the robotics world consists of service robots, a heterogeneous category including all that is used outside manufacturing plants. Here you find drones, the sector that has been growing more quickly in recent years and is now worth around $20 billion.[4] There are legged robots for inspection and maintenance; robots for search and rescue; autonomous transportation vehicles for logistics. Personal assistants, toys, weeding robots for agriculture and milking robots for farming. Surgical robots and exoskeletal robots for rehabilitation and support in physical work. Construction robots. Space rovers. And, of course, software and components for all of the above.

Service robots for domestic tasks are the leading segment in terms of number of units: more than 18 million robotic vacuum cleaners, lawn mowers, and pool and window cleaners were sold in 2020.[5] They were followed by about 600,000 entertainment and educational robots, such as robotic toys, remotely operated drones and cars, mechatronic animals, and small humanoids. Finally, about 131,000 service robots were sold for professional use in mining, agriculture, logistics, and other service sectors. Add to the list a few rovers currently operating on Mars, and you have the full picture.

The field of service robotics is so diverse that mapping all of the leading companies in each subniche would need a chapter on its own. For sure, only one service robotics company so far has truly broken into the mainstream. It is iRobot, the maker of the Roomba vacuum-cleaning robot which is by far the best-selling robot in history. The company's journey, from an MIT spin-off making a living out of DARPA grants to a recognizable global brand, was far from straightforward and says a lot about how crucial inventing a business model along with the technology is for robotics. In a 2006 talk, iRobot's CEO, Colin Angle, listed no fewer than fourteen failed business models they had to try before starting to make vacuum cleaners and sell them to households.[6] They went from "sell movie rights to and perform a robotic mission to the Moon" to "sell research robots to universities." From "earn royalties on robotic toys" to "develop and license technology for nanorobots to clean blood vessels"—two business models that failed for them, but could work for the protagonists of two of our futuristic chapters. They considered selling all sorts of robots to all sorts of industries: inspection robots for power plants, educational robots for museums, land-mine-clearing robots to armies in war zones, mobile robots to data centers that could be controlled over the Internet. They also considered betting on software—the way Microsoft did in the 1980s, ultimately taking the computer business from the hands of hardware maker IBM—and licensing a robot operating system. "For a long time the robot industry was unfundable," Angle later recalled in a blog post commenting on that list of failures.[7] "Why? Because no robot company had a business model worth funding. It turns out that a business model is as important as the tech but much more rarely found in a robot company."

Another company that has found a strong business model is Intuitive Surgical, a Silicon Valley company that since the early 2000s has dominated surgical robots with its Da Vinci system. The drone sector has a few big players, beginning with the Chinese companies DJI and Yuneec, the French Parrot group, and the defense and aerospace American giants like Lockheed Martin and Boeing. Boston Dynamics, a US East Coast company that started out as an MIT spin-off on research contracts with the US Defense Department, is now probably the most recognizable brand for quadruped and humanoid robots. Once part of Google's round of robotics acquisitions, it belonged for a while to the Japanese SoftBank Group before being acquired by Hyundai in 2021.

Each market niche has its own players, and as is often the case for innovative markets, companies appear and disappear almost daily; they grow quickly and then fall, are acquired by larger groups, and then sold to other groups a few years later.

It is here that the really new stuff is going on, and it is here that there is the most room for growth. Andra Keay, founder of the nonprofit association Silicon Valley Robotics that helps US start-ups grow a business around robotics technology, uses the term "Robotics 2.0," as opposed to the "1.0" version that is the Old Guard industrial robotics, saying, "Robotics 2.0 is quantitatively and qualitatively different from what was there before. It is grounded in the combination of sensors and smartness, and the ability to navigate in the real world in real time separates these new products from the established robotics industry. You see fringes of this appearing in the established industry in the form of co-bots or autonomous mobile robots, but most of it is in new companies developing service robots."

As a whole, service robotics is growing probably much more than the numbers themselves show. The prepandemic year, 2019, may not have been a good one for industrial machines, but it was an excellent one for service robots, which grew 32 percent up to more than $11 billion in revenue. But Keay notes that at this stage, revenues tell only a small part of the story: "A lot of the new companies have no revenues yet, but they have investments. If we look at the overall value of investments, mergers, and acquisitions in the robotics business, it is probably around $50 billion per year, which is more than the total revenues for robots and peripherals. That is the hidden value of robotics." Once that hidden value starts to turn into actual value, Robotics 2.0 will enter a new era and will find itself at several crossroads. Let's explore the main ones.

HARDWARE OR SOFTWARE?

Digital technology markets—and robotics is no exception—have a hardware and a software side, which can be separated or intertwined at different degrees, and either of which can be the driving force in shaping the market. The classic structure of the music industry, for example, had a clear separation: consumer electronics companies produced hi-fi sets or CD players and record companies produced recorded music. The two businesses needed each other, but each tended to its own job.

The mobile communication industry started out as a somewhat different affair: Motorola, Nokia, and Ericsson, which dominated its first phase, made the phones and wrote the software to run on them, and so did (and still does) Apple. Later, once iOS and Android had established themselves as standard operating systems, app developers created a flourishing software business leveraging the hardware business. The two markets now feed each other. Smartphones remain objects of desire in their own right, as the aggressive marketing campaigns around each new iteration of the iPhone or the

Samsung Galaxy show. Software developers make the most of constantly evolving screens and processors, creating more intriguing and more demanding apps, which in turn drive users to upgrade their hardware.

Robotics involves both hardware and software too, and the coevolution, divergence, or even power struggle between these two parts of the business will probably have a big impact in shaping the market. As of today, most industrial robotics is hardware driven, and most manufacturers bundle proprietary software with their robots. But things are changing, especially for service robotics: over the last decade, the open source software environment ROS (Robot Operating System) has become a de facto standard, making robots more interoperable. Meanwhile, the costs of robot hardware components (batteries, position sensors, cameras, CPUs) has dramatically gone down. This may ring a bell to those who remember the major transformation of the computer industry when Microsoft established its dominance on the market for operating systems and personal computers became small and inexpensive enough to appeal to everyday consumers. Computer hardware became a commodity and market power shifted from IBM toward software companies such as Microsoft and Oracle. However, at least one company, Apple, insisted on doing both hardware and software and was vindicated in the end.

Will robotics also divide in two businesses, with some companies producing robots and other ones developing software, even downloadable apps? Could the latter in the end become a richer market than the former, as it happened with IBM and Microsoft? According to Andra Keay, there are signs that this is happening. "I see more and more start-ups in robotics whose job is really to deploy software. How many arms with suction grips can there be after all? Even though many companies are providing the full stack, the real competitive advantage is in the algorithm, in computer vision, in machine learning from placement and grasping," she notes.

Keenan Wryobek knows a thing or two about robotics software, having cofounded Willow Garage, the company that invented ROS, with Eric Berger when they were both at Stanford in the mid-2000s. He has a more nuanced view of how things will play out: "I think for the next ten years, we're going to see software companies writing software for really niche applications. They will take something like an off-the-shelf robot arm, use ROS as an operating system, and write a piece of software for it. Robotics software now is basically AI in the real world, and that does not generalize well, so it's very difficult to reuse existing software for new situations." Because of that, Wryobek says, there is a lot of room for software-focused companies that write software for each new application and do not have the time and resources to also work on the hardware, which they would rather buy on the market. But things will change, Wryobek says, when we begin to see breakout applications in the consumer space, comparable to iRobot's big success with Roomba: "When the volume of the market becomes much higher, that will start to justify bigger bets where people develop hardware that is tightly coupled to software." He sees Apple as a good example of what top robotics companies may look like—and work—when the sector is really established: "What they've shown is the value you produce when you bring hardware and software design together, from a quality and reliability point of view. When I think of robotics, especially consumer-facing robotics, the trust and reliability of the hardware is no small thing. I don't see robotics hardware being fully commoditized as computers; it's far harder."

PRODUCT OR SERVICE?

Whether your business is robots, musical instruments, cars, or almost any other product you can think of, the most obvious way to act on the market is to sell the product itself. You make it, you price it, you advertise it, and then it changes hands and becomes the property of the buyer. But in many cases—especially with products that are very expensive, take a lot of space, or require specific skills—there are other possibilities. Cars and pianos can be rented or leased: the customer pays to use them for a limited amount of time and returns them at the end. Or the customer can buy the full service, as when taking a taxi (car as a service) or hiring a keyboard player who will bring her own instrument and play it.

In industrial robotics, the typical way to make business is to sell the machines and maintenance services. The problem is that robots are, with few exceptions, very expensive. A carmaker can buy hundreds of them, but a small to medium enterprise that wants to automate part of its production may not afford even a single one. The same is true for service robots in agriculture. It remains to be seen whether humanoids will develop into a large market—but if so, they surely will not be cheap. On the other hand, until there is enough demand for robots to increase production, economies of scale will not be able to keep prices low.

That explains why an idea called "robotics-as-a-service" has been gaining momentum in the robotics business, often hailed as the future of the industry in specialized publications. Instead of buying a robot and bothering to install and operate it, you can contract a company to provide the service that the robot makes possible. The company owns the robot and the expertise needed to operate it, and you simply enjoy the final result. The ". . . as a service" concept is also gaining popularity in other industries where it may erode traditional ownership. For example, in the car industry, analysts expect shared mobility—where people jump on and off cars from fleets managed by ride-booking apps—to be an integral part of the future, together with electric powertrain and autonomous driving. If we go toward a future where we buy less stuff and share appliances with others, might that apply to robots as well?

Frank Tobe, founder of the Robot Report—a website that tracks and analyzes the evolution of the robotics industry—and a veteran consultant to robotics businesses in the United States and beyond, agrees that the robotics-as-a-service model may take hold in some industries that need to concentrate a lot of machine work in limited time. "Agriculture is one example," he says. "Imagine a monster farm, like the ones they have in Brazil, that have ten days to complete a job. They may need one hundred autonomous tractors, and then they don't need them anymore. The tractors can then move to Chile, or Argentina, or Australia, because you can't let one hundred tractors sit around." The best thing for the farmer would indeed be to contract a company to bring the tractors to the farm, operate them, deliver the harvest, and then say good-bye until the next year—as opposed to buying one hundred tractors and letting them sit in barns for months. "The same thing happens in construction, where you do one big project and then you move to another one," he notes.

Andra Keay mentions the example of Dishcraft Robotics, a California company that provides a robot dishwasher for commercial kitchens that can automatically stack dishes and uses computer vision to recognize dirt and debris on dishware. "For the first trial, they babysat the machine in the kitchen," she recalls. They carefully priced it so that it cost less than two regular dishwashers plus the person tending to them, providing a more cost-effective use of space in the kitchen. But very rapidly, Keay says, the company started to wonder whether they really needed to have the machine in the kitchen at all. They looked at what happens with laundry, another daily need of the restaurant business, where a successful business model has existed for decades. Restaurants and hotels put out the bags with dirty laundry each night, and the cleaning company picks them up and provides clean linens in return, with no washing happening on site. "So Dishcraft became a hub for providing clean dishes as a service. They pick up the dirty ones and replace them with crates of clean dishes, and keep the machines at their premises. It is still possible that large institutional kitchens will want to invest in the machine, but the majority probably won't," she summarizes.

Still, selling the whole product will remain the dominant model, Tobe thinks: "The traditional role of robotics as a service is to make it possible to enter the market. But once the concept is proven, you want your own control and economy of scale." Outside of very innovative technology, Tobe thinks, "this model may stick for the construction or agriculture sectors. But I haven't seen a convincing case yet for a lasting robotics-as-a-service model outside those fields, and I don't think it can happen in the consumer sector."

In summary, expect most future robotics companies to sell their robots directly to customers, except for the new technologies that will periodically emerge, where selling the service may remain a good way to enter the market and test the waters.

MAJOR OR INDEPENDENT?

Many technology markets in the twentieth and early twenty-first centuries have gone through the same trajectory. First, grassroots innovation is brought by many small companies furiously competing with each other and scrambling to stay afloat; then a few winners emerge, absorb most of the competition, and establish an oligopoly—or in some cases, even a monopoly—as the business consolidates and the market grows; a period of market stability ensues, until crucial innovation in technology or business model enables a new emerging competitor to rise, threaten the dominant players, and maybe replace them. The last step can be repeated ad libitum. This is what happened to the oil industry: from the first pioneering steps in the late nineteenth and early twentieth centuries when small entrepreneurs were digging hundreds of wells in Texas and other US states; to the era of European and American majors such as Shell, Texaco, and Standard Oil that dominated the market up to the 1960s; followed by the rise of the OPEC states, which were in turn challenged by the new wave of fracking. This trajectory also applies to the computer industry, where a now largely forgotten early wave of innovation in electronic components in the late nineteenth and early twentieth centuries culminated in the founding of IBM in 1911. The company would become almost synonymous with the computer industry for decades after World War II, until personal computers and new operating systems transformed the market, creating new dominant players. These were then challenged by new emerging competitors with the Web, cloud computing, and mobile computing. Still, the degree of concentration (one, five, ten dominant companies) and the pace of the stability and disruption cycles can vary a lot. The carmaking landscape did not change much for more than fifty years until Tesla began to give headaches to Volkswagen and Toyota and the others. In contrast, the computing industry has seen more or less one revolution per decade.

According to Andra Keay, recent history provides some hints on what could come next for the robotics industry in terms of concentration or diversification. The most successful companies that broke new ground in the robotics market in the last couple of decades have so far managed to stay independent, without being acquired by either existing robotics majors or, for that matter, large multinationals with stakes on those markets. iRobot, for example, is the only robot company to score a blockbuster product on the consumer market. Many have copied it, but no one has bought it. "It took them a long time to get a robot vacuum cleaner accepted in the market,"

Keay recalls. "But they are now the leaders in a new product category that they created, and every major appliance company now has a robot vacuum cleaner in their catalog."

Another example is Intuitive, which established dominance in the surgical market in the early 2000s. "Because many of their patents have recently expired, we are now seeing a proliferation of other products. But interestingly, we are seeing a proliferation of new surgeries that require different techniques and technologies: knee surgery, brain drills, lung cancer," Keay notes. Rather than competing head-to-head with Intuitive on its turf (noninvasive surgery of the abdomen), these companies are trying to conquer new turfs in medicine and establish dominance there. Because robots are not a single product category—like computers, for example—but will result in dozens of entirely new product categories, Keay sees a lot of space for diversification rather than concentration: "It takes a lot for a company to nail the right product at the right time for the right market, and emerge as the leader of a new category. Once that happens, it provokes a ripple up of investments, but the inventor maintains sufficient market leadership to stay independent. For some time at least, it is more likely that big players inside each market sector start doing their own version of the robot product rather than rush to acquire the successful start-up."

In the long term, though, Frank Tobe is less optimistic about the possibility for native robotics companies to remain dominant. "It's sort of inevitable that big industrial players are going to absorb service robotics in their sectors at some point" he says. "But it's not going to be one or two global robotics conglomerates," he says, dismissing the Asimovian scenario of a robotics monopoly. "The big companies within sectors such as construction, health care, aerospace and defense, logistics, and agritech will take over in their respective sectors." Hyundai's acquisition of Boston Dynamics in 2021, signaling the importance of autonomous robotics for the future of mobility, is already a good example of that trend. The most likely future for the robotics business in this respect may be a pendulum constantly swinging between concentration and diversification. On one side, powerful multinationals, not originally robotics companies, take over in highly mature markets where innovation slows down, and on the other side, new companies that periodically appear and grow around entirely new products, creating new markets where they remain leaders as long as the ability to innovate remains key.

WEST, EAST, NORTH, OR SOUTH?

Compared to other high-tech industries, industrial robotics is surprisingly not USA-centric. The United States is—depending on the year you consider—the third or fourth country in the world in terms of annual new installations of robots, after China and Japan, and competes with South Korea for third place. But, as the IFR Industrial Robotics Report notes, "Most of the robots in the USA are imported from Japan, Korea, and Europe. There are not that many North American robot manufacturers."[8] Although the first industrial robot, Devol and Engelberger's Unimate, was born in the United States, Asian and European companies have dominated the market for decades.

With the rise of service and consumer robotics, Andra Keay sees a "power shift" toward the United States, though still largely unnoticed. "There is more robotics in Silicon Valley than anywhere else in the world, put together," she says. "But it is a sort of secret, because Silicon Valley is not home to any of the major industrial players, and the majority of companies there have no revenues yet." But they have investments. In 2010 there was practically no investment in robotics. "By 2015 I had tracked a billion dollars of investments, and in hindsight it was too conservative; it was probably closer to $3 billion. Globally, we are now probably at $30 billion, for the largest share in Silicon Valley, followed by China and by the Boston–New York area."

The United States is now leading in innovation, she says. "Europe understands the direction it needs to go, but is too slow." As for China, "There's a lot you can do if you decide to do the things others are already doing, but you do it locally in a huge local market. India is also growing, in part thanks to the US immigration policy. There are thousands of Indian engineers who have spent twenty years in Silicon Valley and are still waiting for permanent residency. They will be taking their skills and their money back to India and help develop a healthy industry there."

The new cold war between the United States and China is also sending ripples through the business. "Up to a few years ago," Keay notes, "many innovative robotics companies in the United States were being funded by investment groups that were connected to China. When the US Defense Department realized that, it shut the door to all China-backed investments, causing a migration of investments to other countries such as Israel, which plays a neutral geopolitical game and whose companies are very good at bridging the gap from innovation to market."

Hints of future geopolitics of robotics and AI may be gained by technical publications and patent activities. According to a 2019 report on AI by the World Intellectual Property Organization, China and the United States dominate research on AI and robotics in terms of number of scientific publications and patents.[9] While most patent requests were filed in the US Patent Office—about 150,000 patents up to 2019—the China patent office came close with almost 140,000 patents. There is one important difference: the majority of filings in China are made by Chinese applicants, whose filings have increased exponentially in recent years, while many patents filed in the United States are secondary filings by inventors who already applied for European or Asian patents. Furthermore, China leads in the number of AI patents for robotics. However, the United States ranks first in terms of highly cited patents, which may reflect the higher impact or older status of US patents. While companies dominate patent activities and the top twenty companies filing AI-related patents are from Japan, the United States, and China, universities play a leading role in emerging application fields, including robotics, and seventeen of the top twenty universities filing AI-related patents are Chinese. These data suggest that industrial power in intelligent robotics will concentrate in the United States and China, with Europe's role being gradually diminished and new players such as Israel and India taking up shares of the market.

And then there's Africa, currently almost absent from the robotics map but with the potential to become a huge market for robotics and to nourish new businesses—if not robotic manufacturers, surely providers of services based on robots. "Opportunities there are endless," says Sonja Betschart, the cofounder of WeRobotics, a nonprofit organization that supports training on robotics and helps the growth of robotics ecosystems in Asia, Africa, and Latin America. "For example, in Africa there is a growing market for farm robotics and smart framing to cover the growing demand on food systems all the while addressing limited natural resources for growth. This is where robots of any kind find their integrated role," Betschart notes. "Drones can be used to map fields to identify crop health or contribute with data to irrigation decisions. Other types of drones or wheeled robots can then apply fertilizers or treat issues in a very precise way." Another example is connectivity: most connectivity in Africa now comes from mobile networks—hence, an important network of cell phone towers that needs periodic monitoring and inspection that can be addressed with drones. Still, she notes that a problem that holds back robotics applications in Africa, in addition to cost, is that "the countries that produce drones and robots today design and build their systems to fit their own markets. Most robotic systems are very closed, difficult to repair, and fit conditions in the West but do not fit the conditions in Africa, the South Pacific or Latin America, from weather to bandwidth to access to charging capacity." In contrast to this, she notes that "most problems that afflict countries are very local problems that need to be addressed by local experts, adapting the technology to their specific needs and not the other way around." Still, Betschart sees a lot of opportunities for African businesses in applications and added services: "It's a bit like cars: while not many countries manufacture cars, they allow creating economic opportunities everywhere, as long as they can be adapted to their environments, and are locally repairable."

A good example of someone who built a pioneering robotics application in Africa comes from Keenan Wryobek. A few years after creating ROS, he cofounded Zipline, a company that uses autonomous drones to deliver blood, vaccines, and other medical supplies to hospitals in Rwanda and Ghana, and now in the United States, a successful venture that to this day is the poster child of flying robots' potential in the developing world. "In Africa there is a huge appetite for trying new things," he says. "Across all industries, from telemedicine to electrification, there is an amazing marketplace where good products are winning, people are adopting them, and industries are leapfrogging what we are doing in the West. People always talk about the success story of mobile phones in Africa, but that is just the tip of the iceberg." And although he agrees that the future of Africa will mostly be about creating businesses based on robots built elsewhere, Wrjobek sees signs of a production capacity thanks to better access to supply chains and better education. For example, his company recently started manufacturing drone battery packs on site in Rwanda (and battery packs are going to be a big business in the future electrified world, as Tesla's bold plans in this field show).

Few have spent as much energy trying to bring robotics to Africa as Jonathan Ledgard, a journalist and novelist who was for some time the director of Afrotech, an EPFL think-tank on new technologies for Africa. Having witnessed firsthand, as East Africa correspondent for the *Economist*, the transformative impact of mobile phones in the continent, he became convinced that robots—and drones in particular—could do for physical goods what mobile phones had done for information goods: they could help public services and businesses circumvent the obstacles caused by poor infrastructure on the ground (landlines for mobile phones, roads for robots) and leapfrog directly to next-generation technology instead of waiting for adoption of twentieth-century technology. He spent many years working with African governments, nongovernmental organizations, and international organizations such as the World Bank to promote visionary projects, such as a network of droneports (triggering the involvement of no less than the renowned architect Norman Foster) and helping robotics companies create a business there. Among other things, he helped Zipline start operations in Rwanda.

Ledgard is convinced that robotics will play a big role in transforming the economies of many African countries due to a few converging trends. "One is urbanization; that means that simply going or moving something from A to B is increasingly a problem. What is interesting about drones, more than drones themselves, is that they are a use case for affordable robots; they're the first robots that make sense for many people." But once walking, legged, or swimming robots begin to equally "make sense" (in other words, become affordable and easy to use enough for individuals and organizations to apply them in everyday life) they will prove just as promising for urban Africa. And the most interesting things, Ledgard says, may happen not in megacities but in towns: those that now have ten or twenty thousand people but will get to eighty thousand in a couple of decades. "I think in those towns you will really see a whole range of robotics moving and doing things."

Another important trend is climate change, which may end up accelerating the diffusion of robotics in Africa. Industrial and oil-dependent countries such as Saudi Arabia, he notes, will have to transfer "extremely large sums of money" for mitigation and reparations to the South of the world in order to stay in the carbon business. "I would argue it's a no-brainer that a sliver of that money should go to robotics: R&D investments, education investments, capitalization of existing or new companies." Add to that the new cold war between China, which heavily invested in Africa over the past decade, and the United States that is determined to regain influence there. "Because Africa is being contested, things are going to be very interesting, and if I want to be very cynical, I would say that if you are a Western robotic company you can work this to your own advantage. There are many reasons—humanitarian, strategic, security profit-driven—to be in this enormously growing market."

The result, according to Ledgard, is not necessarily that Africa will produce thousands of robots but that businesses will flourish around robot-based services—particularly for agriculture, mobility, and energy—and adaptation and repair of robots. "There is a lot of informal economy across Africa, for example, with motorbike and car repair. Similarly, I imagine little shops in African towns, repairing robots or combining off-the-shelf robotics components to create low-cost robotic widgets to help humans in many tasks." Ledgard notes what is already happening with electronics, where people buy spare components on the marketplace, disassemble and modify computer motherboards, and get them to work on something completely different from what they were designed for. "I imagine a strange future that will look a little like the planet Tattoine in *Star Wars*—a dusty, dirty place with many rooms filled with robots that do tasks and repair things. It still won't be rich, it will be insecure, but there'll be a lot of technology in there helping people do business."

SOONER OR LATER?

Ultimately many answers to the questions we have asked about the future of the robotics industry will depend on the timing of its growth. Will it be slow and gradual, or will it accelerate in a burst due to some sudden technical advancement or breakthrough product that drives the growth of the whole sector? Will it explode sooner, in a world where the US economy still has its leading role and geopolitical influence, or later, in a likely scenario where China has taken over? Will the robotics industry find itself operating in a different kind of capitalism—one, for example, where states take a larger role in guiding technological innovation, the kind of new capitalism that an increasing number of economists and politicians now advocate in order to cope with rising inequality and globalization? These are hard questions for which no one has the answers, although there are signs that the COVID-19 pandemic has accelerated the transitions toward automation in many industries that were previously more cautious about it, such as health care. All we know is that the rest of the twenty-first century will be an incredibly transformative time for the world economy because of opportunity and necessity, and the robotic industry will be part of that transformation.

EPILOGUE: WHAT COULD GO WRONG, OR THE ETHICS OF ROBOTICS

For the most part, this book has painted a somewhat rosy picture of the future of robotics. It reflects a vision of the future that motivates thousands of roboticists around the world. Not all scientists quoted in this book may share the stories of the future narrated in this book, but they all surely have one of their own where their work will have a positive impact on the world.

But we—and the scientists quoted in this book—are well aware that as robots exit research labs, the picture will get more complicated. New technologies solve problems, but often they create new ones in the process. It has happened with almost every important invention in human history—from agriculture to nuclear power—and it is something about which we are reminded every day by global warming, the number one unwanted side effect of industrial technology. Today's scientists cannot be as naive as their predecessors in past centuries were and cannot afford to ignore the possible negative consequences of their work or the societal resistance that their innovations may face.

Roboticists are constantly reminded of possible risks of their work by science-fiction writers who have been depicting dystopian scenarios where robots rebel against humans, conquer the world, and subjugate their former masters. The greatest of all robotics fiction writers, Isaac Asimov, famously wrote three fundamental laws that were supposed to be incorporated in robots' nature to prevent this scenario:

A robot may not injure a human being or, through inaction, allow a human being to come to harm.

A robot must obey the orders given it by human beings except where such orders would conflict with the First Law.

A robot must protect its own existence as long as such protection does not conflict with the First or Second Laws.[1]

The three laws were meant to ensure that intelligent and self-conscious robots would not prioritize themselves above humans, and that they would serve humans rather than themselves or other robots.

Today, most roboticists agree that the discipline needs to develop its own ethical guidelines to make sure that the opportunities it creates are not outweighed by negative effects and that the good it does for most people is not offset by the damage caused to others. But they also agree that those guidelines will have to be more sophisticated than Asimov's three laws. *Roboethics*, a word coined at an international symposium in 2004, now identifies a thriving research field that tries to map areas where future robots may clash with people's safety, needs, and rights, as well as the reasons that may make society hesitant to accept intelligent machines.[2] The field is less concerned than Asimov was about the ethics of robots (giving a sense of morality to imaginary conscious machines that are still nowhere to be seen) and more about the ethics of roboticists: making sure that human beings make ethical choices when designing, building, selling, and operating robots.

In 2010, for example, a group of European roboticists outlined a set of robotics principles for the UK Engineering and Physical Sciences Research Council (EPSRC) that look like an updated, science-based version of Asimov's laws:

1. Robots should not be designed as weapons, except for national security reasons.
2. Robots should be designed and operated to comply with existing law, including privacy.
3. Robots are products: as with other products, they should be designed to be safe and secure.
4. Robots are manufactured artefacts: the illusion of emotions and intent should not be used to exploit vulnerable users.
5. It should be possible to find out who is responsible for any robot.[3]

While some concerns are as old as robots themselves, others acquire a whole new meaning when referring to the new generation of robots described in this book. First, unlike industrial robots, these machines will enter many more environments beside the workplace and will physically interact with humans more frequently. Second, they are going to be more autonomous and will rely less on classical control methods that make today's machine predictable.

The concern of robots' autonomy has nothing to do with the concern of the robot's consciousness. Few roboethicists today worry about Asimovian scenarios of superintelligent and conscious robots rebelling against their creators. If such machines are possible at all, we are still so far away from creating them that it does not even make sense to be concerned. But roboticists, as well as politicians and citizens, have good reason to be worried about partially autonomous, sloppily designed machines making bad decisions in the proximity of people. There are many concerns—for example, that autonomous vehicles may cause fatal accidents because of bad decisions based on wrong readings of their sensor signals. No technology is born perfect. But all technologies evolve and become safer and more reliable because engineers monitor and analyze mistakes. If something goes wrong with a robot, we'll need to know where it went wrong exactly and which design choice allowed it to happen. And we must be able to assign blame so that robot designers have an incentive to do things right and think of the consequences. That's how aircraft, for example, became safer, and we want the same thing for robots.

"People often talk about robots being accountable," notes Alan Winfield, a specialist in cognitive robotics who is now professor of robot ethics at the University of Bristol and who was part of the working group behind the five EPSRC principles. "But robots cannot be accountable. For me, the real question is: How do we make designers, manufacturers, operators accountable? The prerequisite of that is the transparency of the system. We should be able to trace and explain the course of the decisions a robot makes," Winfield says. But that explanation can get tricky when modern machine learning methods are involved—as is the case for most of the robots, existing and imaginary, that we have described in this book. In the kind of deep learning that is increasingly used in intelligent robots, decisions emerge through several rounds of trial and error from the inner workings of neural networks in ways that are often somewhat mysterious also for their designers.

"Robotics will inevitably inherit the problems of machine learning," says Winfield. "And bias is possibly the most egregious of those problems." We already know that facial recognition systems, automatic translators or automatic text generators—currently some of AI's most successful applications—can display worrying biases such as not recognizing nonwhite faces, or arbitrarily making nouns feminine or masculine in a translation based on the profession of the person. When those machine learning techniques are used on robots that have to physically interact with people, the problem can be even greater. Winfield quotes an example from his own lab where researchers worked with a group of children using a toy robot. "One of the robot's features was the ability to learn and recognize faces of children. The robot successfully learned to recognize all but one child in the group, and that caused a lot of distress to that child. And the problem is we do not know why it failed." Extend this example to all the situations where social robots will interact with humans in the future, and you'll see the magnitude of the problem. Glitches in the data used by machine learning could cause robots to rely on stereotypes, neglect minorities, or favor users from specific ethnic or cultural backgrounds. And in the very worst case, the failure on the robot's part to properly interpret a situation may result in physical dangers for humans—as when the sensors of an autonomous vehicle fail to recognize a pedestrian in that shadow on the side of the road.

Winfield suggests a couple of solutions. "In order to avoid bias, one thing we'll need soon is more curated data sets," he says. "The problem with current AI is that big data sets are just scraped from the Internet, and that is asking for trouble. We need to develop methods to create data sets live from the real world—for example, for grasping and social interactions—instead of relying on indirect data such as footage available on the Internet—and we need to curate them." More in general, he suggests that robotics should beware of pure, model-free machine learning and use a hybrid of machine learning and "good old-fashioned algorithmic AI," where the system does not really learn from scratch but is given models of itself and the world, representations that allow the robot to better predict the consequences of its actions and act as boundaries to the bottom-up learning process, by injecting "notions" in the robot that it would not discover by machine learning alone.[4] This idea is part of a broader debate in the AI community on whether deep learning techniques can really get to grasp the complexity of the world by learning it from the data alone, without previous notions encoded by humans. Many scientists argue for the need to inject machine learning with physical models of how objects and forces behave, which, for example, could be used to encode safety rules. Winfield believes something similar should be done with psychological, emotional, and ethical models to prevent bias.

Winfield also suggests that robots—in particular, social robots that work among people—should have an "ethical black box, the equivalent of an aircraft's flight data recorder (which, despite most people calling it black, is normally orange)."[5] This would continuously record sensor and other internal status data, possibly documenting the AI decision-making process. The data could be erased periodically as long as all works fine, but could be downloaded and analyzed in case of a serious accident or of an evidence-biased behavior. The presence of this box, and of a system in place where professionals and dedicated institutions systematically analyze when something goes wrong, would increase trust among users and make robots more reliable.

While bias, models, and black boxes have to do with how robots "see" humans and the world, another set of problems has to do with how humans see robots—in particular, machines that sense the world, display a good degree of autonomy, and can, to a certain extent, resemble living things. People may unconsciously perceive them as alive, develop attachment and unrealistic expectations, and even become addicted.

"A key principle must be not to design robots in a way that deceives people into thinking they are persons," says Winfield. We humans are prone to anthropomorphization, we know that happens even if the robot is not anthropomorphic, and designers should make efforts to minimize that effect. This is particularly important for safeguarding weak people."

Joanna Bryson, a professor of ethics and technology at the Hertie School in Berlin, has noted, for example, that

> we could in principle design interactive AI systems to "suffer" if humans do not pay attention, but that would be an intentional and unnecessary design choice. . . . Robots should not be designed in a deceptive way to exploit vulnerable users; instead their machine nature should be transparent. Transparent here does not mean "open source"; while often helpful, that is neither necessary nor sufficient. Rather, transparency implies comprehensibility. One of the most basic ways to keep AI transparent is to avoid making it appear human. Robocalls should sound like robots. . . . Robots, even sex robots, should be evidently mechanical so that we don't come to associate strict subordinance—indeed, ownership—with actual interhuman relations.[6]

As Kate Darling puts it, "What keeps me up at night isn't whether a sex robot will replace your partner; it's whether the company that makes the sex robot will exploit you."[7]

Toby Walsh, a professor of artificial intelligence at the University of New South Wales, has suggested that robots should have a "Turing red flag," using an analogy from the early days of motorization, when most people were unaccustomed to the risk of being run over by a car.[8] "Concerned about the impact of motor vehicles on public safety, the U.K. parliament passed the Locomotive Act in 1865. This required a person to walk in front of any motorized vehicle with a red flag to signal the oncoming danger." By analogy, he suggests that "an autonomous system should be designed so that it is unlikely to be mistaken for anything besides an autonomous system, and should identify itself at the start of any interaction with another agent." This would apply to software—when interacting with customer service, for example, you should always know whether you are chatting with a real person or with an AI bot—and to hardware—for example, Walsh suggests that autonomous cars should have different registration plates from crewed ones that clearly mark them as such.

Finding technological solutions is not necessarily the most difficult aspect of roboethics. Assuming those solutions work, the real problem is making them happen. Winfield is convinced that robotics will soon need standards—common design rules, requirements, and measurable performance levels that all designers and manufacturers have to stick to. In a field that is still at an early stage and where a lot of innovation has yet to happen, the idea of governments imposing standards by law irks many scientists. It is the case of the European Commission's 2021 sweeping proposal for AI regulation, to which members of CLAIRE (a confederation of AI European laboratories) have responded with a number of concerns, such as an excessive vagueness on the definition of citizens' rights that would be affected by AI and the risk that such regulation may erode European competitiveness in the field. In this case, as in many other ones, what we are seeing with software-level AI is likely only an anticipation of what will happen for physical AI.

"But I believe it's a myth that regulation stifles innovation," Winfield says. "On the contrary, well-written regulations provide a framework for innovation and enable it, as we've seen in many industries. And standards do not even need to be written into law; most of them are voluntary," he notes. "If governments are procuring robotics and AI systems, for example, they can and should require their suppliers to comply with certain standards—which could include the ethical black box—as part of the conditions for the supply. Requirements for prime government contractors then trickle down along the supply chain." As Winfield noted in a 2019 article, licensing authorities, professional bodies, and governments can all use "soft governance" to influence the adoption of standards.[9] And in a competitive mass market for robots, such as the one we might have one day, compliance with ethical standards can translate into market advantage.

Roboethics has only begun to scratch the surface of the challenges that a new generation of robots, which borrow cognitive and physical abilities from living things, would pose to society. Other issues will come up along with new technological achievements. Ethicists, legislators, and scientists themselves will struggle to keep up—and so will the general public. Perhaps it's a good thing that robotics is so much harder than software: unlike what is happening with some applications of machine learning, with robots we do have time to investigate their consequences and adjust the direction of innovation. But that will happen only if robotics becomes a collective endeavor, where there is a constant interchange between what happens in laboratories and what happens in society and where people participate in innovation rather than passively receive it. Of all the challenges presented in this book, this may be the most difficult one.

Acknowledgments

Several scientists and experts spent precious time sharing with us their vision of the future. Whether or not they were explicitly quoted, they all helped us shape this book: Lucia Beccai, Sonja Bertschart, Marina Bill, Aude Billard, Josh Bongard, Oliver Brock, Grégoire Courtine, Marco Hutter, Auke Ijspeert, Andra Keay, Mirko Kovac, Rebecca Kramer-Bottiglio, Rafael Lalive, Cecilia Laschi, Hod Lipson, Barbara Mazzolai, Arianna Menciassi, Robin Murphy, Radhika Nagpal, Brad Nelson, Anibal Ollero, Fiorenzo Omenetto, Daniela Rus, Davide Scaramuzza, Thomas Schmickl, Metin Sitti, Frank Tobe, Conor Walsh, Alan Winfield, and Keenan Wryobek.

Alessio Tommasetti created the illustrations. We hope that they will help readers get a sense of the future we had in mind when telling our fictional stories, but also leave them free to imagine their own version of that future.

Finally, a big thank to Marie L. Lee and Elizabeth P. Swayze at MIT Press for respectively encouraging the writing of this book and ensuring its completion in due time and size and to the anonymous reviewers for constructive and helpful feedback on various drafts of the manuscript.

Notes

INTRODUCTION

1. Bill Gates, "A Robot in Every Home," *Scientific American*, February 1, 2008, https://www.scientificamerican.com/article/a-robot-in-every-home-2008-02/.

2. International Federation of Robotics, "IFR Industrial Robotics Report" (International Federation of Robotics, 2021).

3. International Federation of Robotics, "IFR Service Robotics Report" (International Federation of Robotics, 2021).

4. International Federation of Robotics, "IFR Industrial Robotics Report."

CHAPTER 1

1. Wolfgang Cramer et al., "Climate Change and Interconnected Risks to Sustainable Development in the Mediterranean," *Nature Climate Change* 8, no. 11 (2018): 972, https://doi.org/10.1038/s41558-018-0299-2.

2. "Venice Tests Its Mighty Flood Defences," *Economist*, August 20, 2020, https://www.economist.com/europe/2020/07/16/venice-tests-its-mighty-flood-defences.

3. Tamás Vicsek, András Czirók, Eshel Ben-Jacob, Inon Cohen, and Ofer Shoche, "Novel Type of Phase Transition in a System of Self-Driven Particles," *Physical Review Letters* 75, no. 6 (1995): 1226–1229, https://doi.org/10.1103/PhysRevLett.75.1226.

4. Ronald Thenius et al., "SubCULTron—Cultural Development as a Tool in Underwater Robotics," in *Artificial Life and Intelligent Agents*, ed. Peter R. Lewis Christopher J. Headleand, Steve Battle, and Panagiotis D. Ritsos (Berlin: Springer, 2018), 27–41.

5. Robert Kwiatkowski and Hod Lipson, "Task-Agnostic Self-Modeling Machines," *Science Robotics* 4, no. 26 (2019), https://doi.org/10.1126/scirobotics.aau9354.

6. Shuguang Li et al., "Particle Robotics Based on Statistical Mechanics of Loosely Coupled Components," *Nature* 567, no. 7748 (2019): 361–365, https://doi.org/10.1038/s41586-019-1022-9.

CHAPTER 2

1. Chris Goldfinger et al., "Turbidite Event History—Methods and Implications for Holocene Paleoseismicity of the Cascadia Subduction Zone," professional paper 1661-F (Reston, VA: US Geological Survey, 2012), https://doi.org/10.3133/pp1661F.

2. Goldfinger et al., "Turbidite Event History."

3. Robin Murphy, *Disaster Robotics* (Cambridge, MA: MIT Press, 2014).

4. Stefano Mintchev and Dario Floreano, "Adaptive Morphology: A Design Principle for Multimodal and Multifunctional Robots," *IEEE Robotics Automation Magazine* 23, no. 3 (2016): 42–54, https://doi.org/10.1109/MRA.2016.2580593.

5. Ludovic Daler, Stefano Mintchev, Cesare Stefanini, and Dario Floreano, "A Bioinspired Multi-Modal Flying and Walking Robot," *Bioinspiration and Biomimetics* 10, no. 1 (2015): 016005, https://doi.org/10.1088/1748-3190/10/1/016005.

6. Stefano Mintchev and Dario Floreano, "A Multi-Modal Hovering and Terrestrial Robot with Adaptive Morphology," in *Undefined* , Philadelphia, 2018, https://infoscience.epfl.ch /record/255681.

7. Ajanic, Enrico, Mir Feroskhan, Stefano Mintchev, Flavio Noca, and Dario Floreano. "Bioinspired Wing and Tail Morphing Extends Drone Flight Capabilities." Science Robotics 5, no. 47 (October 28, 2020): eabc2897. https://doi.org/10.1126/scirobotics.abc2897.

8. Davide Falanga, Kevin Kleber, Stefano Mintchev, Dario Floreano and Davide Scaramuzza, "The Foldable Drone: A Morphing Quadrotor That Can Squeeze and Fly," in *IEEE Robotics and Automation Letters* 4, no. 2 (2019): 209–216, https://doi.org/10.1109/ LRA.2018.2885575.

9. Jemin Hwangbo, Joonho Lee, Alexey Dosovitskiy, Dario Bellicoso, Vassilios Tsounis, Vladlen Koltun, and Marco Hutter, "Learning Agile and Dynamic Motor Skills for Legged Robots," *Science Robotics* 4, no. 26 (2019), https://doi.org/10.1126/scirobotics.aau5872; Joonho Lee, Jemin Hwangbo, Lorenz Wellhausen, Vladlen Koltun, and Marco Hutter, "Learning Quadrupedal Locomotion over Challenging Terrain," *Science Robotics* 5, no. 47 (2020): eabc5986, https://doi.org/10.1126/scirobotics.abc5986.

10. Lee et al., "Learning Quadrupedal Locomotion."

11. K. Karakasiliotis et al., "From Cineradiography to Biorobots: An Approach for Designing Robots to Emulate and Study Animal Locomotion," *Journal of the Royal Society Interface* 13, no. 119 (2016): 20151089, https://doi.org/10.1098/rsif.2015.1089.

12. Murphy, *Disaster Robotics*.

13. Joshua Michael Peschel and Robin Roberson Murphy, "On the Human–Machine Interaction of Unmanned Aerial System Mission Specialists," in *IEEE Transactions on Human-Machine Systems* 43, no. 1 (2013): 53–62, https://doi.org/10.1109/TSMCC.2012.2220133.

14. Boris Gromov, Gabriele Abbate, Luca Maria Gambardella, and Alessandro Giusti, "Proximity Human-Robot Interaction Using Pointing Gestures and a Wrist-Mounted IMU," in *Proceedings of the 2019 International Conference on Robotics and Automation*, 2019, 8084–8091, https://doi.org/10.1109/ICRA.2019.8794399.

15. Jenifer Miehlbradt, A. Cherpillod, S. Mintchev, M. Coscia, F. Artoni, D. Floreano, and S. Micera, "Data-Driven Body-Machine Interface for the Accurate Control of Drones," *Proceedings of the National Academy of Sciences* 115, no. 31 (2018): 7913–7918, https://doi.org/10.1073/pnas.1718648115.

16. Carine Rognon, S. Mintchev, F. Dell'Agnola, A. Cherpillod, D. Atienza, and D. Floreano, "FlyJacket: An Upper Body Soft Exoskeleton for Immersive Drone Control," *IEEE Robotics and Automation Letters* 3, no. 3 (2018): 2362–2369, https://doi.org/10.1109/LRA.2018.2810955.

17. Matteo Macchini, Fabrizio Schiano, and Dario Floreano, "Personalized Telerobotics by Fast Machine Learning of Body-Machine Interfaces," *IEEE Robotics and Automation Letters* 5, no. 1 (2020): 179–186, https://doi.org/10.1109/LRA.2019.2950816.

18. Geert De Cubber, Daniela Doroftei, Daniel Serrano, and Keshav Chintamani, "The EU-ICARUS Project: Developing Assistive Robotic Tools for Search and Rescue Operations," in *Proceedings of the 2013 IEEE International Symposium on Safety, Security, and Rescue Robotics* (Piscataway, NJ: IEEE, 2013), 1–4, https://doi.org/10.1109/SSRR.2013.6719323.

19. Ivana Kruijff-Korbayová et al., "Deployment of Ground and Aerial Robots in Earthquake-Struck Amatrice in Italy (Brief Report)," in *Proceedings of the 2016 IEEE International Symposium on Safety, Security, and Rescue Robotics* (Piscataway, NJ: IEEE, 2016), 278–279, https://doi.org/10.1109/SSRR.2016.7784314.

20. K. Y. Ma, P. Chirarattananon, S. B. Fuller, and R. J. Wood, "Controlled Flight of a Biologically Inspired, Insect-Scale Robot," *Science* 340, no. 6132 (2013): 603–607, https://doi.org/10.1126/science.1231806.

21. Y. Chen, H. Wang, E. F. Helbling, N. T. Jafferis, R. Zufferey, A. Ong, K. Ma, N. Gravish, P. Chirarattananon, M. Kovac, and R. J. Wood, "A Biologically Inspired, Flapping-Wing, Hybrid Aerial-Aquatic Microrobot," *Science Robotics* 2, no. 11 (2017): eaao5619, https://doi.org/10.1126/scirobotics.aao5619.

22. Graule, M. A., P. Chirarattananon, S. B. Fuller, N. T. Jafferis, K. Y. Ma, M. Spenko, R. Kornbluh, and R. J. Wood, "Perching and Takeoff of a Robotic Insect on Overhangs Using Switchable Electrostatic Adhesion," *Science* 352, no. 6288 (2016): 978–982, https://doi.org/10.1126/science.aaf1092.

23. Rebecca McGill, Nak-seung Patrick Hyun, and Robert J. Wood, "Modeling and Control of Flapping-Wing Micro-Aerial Vehicles with Harmonic Sinusoids," in *IEEE Robotics and Automation Letters* 7, no. 2 (2022): 746–753, https://doi.org/10.1109/LRA.2021.3132921.

CHAPTER 3

1. "Zoobotics," *Economist*, July 7, 2011, https://www.economist.com/science-and-technology/2011/07/07/zoobotics.

2. Ali Sadeghi, Alice Tonazzini, Liyana Popova, and Barbara Mazzolai, "A Novel Growing Device Inspired by Plant Root Soil Penetration Behaviors," *PLoS ONE* 9, no. 2 (2014): e90139, https://doi.org/10.1371/journal.pone.0090139.

3. John P. A. Ioannidis, Jeroen Baas, Richard Klavans, and Kevin W. Boyack, "A Standardized Citation Metrics Author Database Annotated for Scientific Field," *PLoS Biology* 17, no. 8 (2019): e3000384, https://doi.org/10.1371/journal.pbio.3000384.

4. "NASA Training 'Swarmie' Robots for Space Mining," *IEEE Spectrum: Technology, Engineering, and Science News*, accessed September 14, 2020, https://spectrum.ieee.org/automaton/robotics/military-robots/nasa-training-swarmie-robots-for-space-mining.

5. J. Scott Turner, *The Extended Organism: The Physiology of Animal-Built Structures* (Cambridge, MA: Harvard University Press, 2009).

6. E. Bonabeau, "A Model for the Emergence of Pillars, Walls and Royal Chambers in Termite Nests," *Philosophical Transactions of the Royal Society of London. Series B: Biological Sciences* 353, no. 1375 (1998): 1561–1576, https://doi.org/10.1098/rstb.1998.0310.

7. Plerre-P. Grassé, "La reconstruction du nid et les coordinations interindividuelles chez Bellicositermes natalensis et Cubitermes sp. la théorie de la stigmergie: Essai d'interprétation du comportement des termites constructeurs," *Insectes Sociaux* 6, no. 1 (1959): 41–80, https://doi.org/10.1007/BF02223791.

8. Justin Werfel, Kirstin Petersen, and Radhika Nagpal, "Designing Collective Behavior in a Termite-Inspired Robot Construction Team," *Science* 343, no. 6172 (2014): 754–758, https://doi.org/10.1126/science.1245842.

9. Michael Rubenstein, Alejandro Cornejo, and Radhika Nagpal, "Programmable Self-Assembly in a Thousand-Robot Swarm," *Science* 345, no. 6198 (2014): 795–799, https://doi.org/10.1126/science.1254295.

10. Meghina Sachdev, "Breakthrough of the Year: The Top 10 Scientific Achievements of 2014," *Science*, December 18, 2014, https://www.science.org/content/article/breakthrough-year-top-10-scientific-achievements-2014.

11. Maira Saboia, Vivek Thangavelu, and Nils Napp, "Autonomous Multi-Material Construction with a Heterogeneous Robot Team," *Robotics and Autonomous Systems* 121 (2019): 103239, https://doi.org/10.1016/j.robot.2019.07.009.

CHAPTER 4

1. D. Floreano and R. Wood, "Science, Technology and the Future of Small Autonomous Drones," *Nature* 521, no. 7553 (2015): 460–466.

2.   Davide Falanga, Kevin Kleber, and Davide Scaramuzza, "Dynamic Obstacle Avoidance for Quadrotors with Event Cameras," *Science Robotics* 5, no. 40 (2020), https://doi.org/10.1126/scirobotics.aaz9712.

3.   Antonio Loquercio, Elia Kaufmann, René Ranftl, Alexey Dosovitskiy, Vladlen Koltun, and Davide Scaramuzza, "Deep Drone Racing: From Simulation to Reality with Domain Randomization," *IEEE Transactions on Robotics* 36, no. 1 2020): 1–14, https://doi.org/10.1109/TRO.2019.2942989.

4.   Kaushik Jayaram and Robert J. Full, "Cockroaches Traverse Crevices, Crawl Rapidly in Confined Spaces, and Inspire a Soft, Legged Robot," *Proceedings of the National Academy of Sciences* 113, no. 8 (2016): E950–957, https://doi.org/10.1073/pnas.1514591113.

5.   Adam Klaptocz, Ludovic Daler, Adrien Briod, Jean-Christophe Zufferey, Dario Floreano, "An Active Uprighting Mechanism for Flying Robots," *IEEE Transactions on Robotics* 28, no. 5 (2012): 1152–1157, https://doi.org/10.1109/TRO.2012.2201309.

6.   Adrien Briod, Przemyslaw Kornatowski, Jean-Christophe Zufferey, and Dario Floreano, "A Collision-Resilient Flying Robot," *Journal of Field Robotics* 31, no. 4 (2014): 496–509, https://doi.org/10.1002/rob.21495.

7.   Andrew M. Mountcastle and Stacey A. Combes, "Biomechanical Strategies for Mitigating Collision Damage in Insect Wings: Structural Design versus Embedded Elastic Materials," *Journal of Experimental Biology* 217, no. 7 (2014): 1108–1115, https://doi.org/10.1242/jeb.092916.

8.   Stefano Mintchev, Jun Shintake, and Dario Floreano, "Bioinspired Dual-Stiffness Origami," *Science Robotics* 3, no. 20 (2018), https://doi.org/10.1126/scirobotics.aau0275.

9.   Joshuah K. Stolaroff, Constantine Samaras, Emma R. O'Neil, Alia Lubers, Alexandra S. Mitchell, and Daniel Ceperley, "Energy Use and Life Cycle Greenhouse Gas Emissions of Drones for Commercial Package Delivery," *Nature Communications* 9, no. 1 (2018): 409, https://doi.org/10.1038/s41467-017-02411-5.

10. Norman Foster Foundation, "Droneport: Rwanda, Africa, 2014–2017," Norman Foster Foundation website, accessed January 25, 2022, https://www.normanfosterfoundation.org/project/droneport/.

11. Przemyslaw Mariusz Kornatowski, Anand Bhaskaran, Grégoire M. Heitz, Stefano Mintchev, and Dario Floreano, "Last-Centimeter Personal Drone Delivery: Field Deployment and User Interaction," *IEEE Robotics and Automation Letters* 3, no. 4 (2018): 3813–3820, https://doi.org/10.1109/LRA.2018.2856282.

12. Przemyslaw Mariusz Kornatowski, Mir Feroskhan, William J. Stewart, and Dario Floreano, "A Morphing Cargo Drone for Safe Flight in Proximity of Humans," *IEEE Robotics and Automation Letters* 5, no. 3 (2020): 4233–4240, https://doi.org/10.1109/LRA.2020.2993757.

13. Naomi LaChance, "Dr. Mirko Kovac: The Drone Optimist," Center for the Study of the Drone at Bard College, January 9, 2015, https://dronecenter.bard.edu/dr-mirko-kovac-the-drone-optimist/.

14. Barrie Dams, Sina Sareh, Ketao Zhang, Paul Shepard, Mirko Kovac, and Richard J. Ball, "Aerial Additive Building Manufacturing: Three-Dimensional Printing of Polymer Structures Using Drones," *Proceedings of the Institution of Civil Engineers—Construction Materials* 173, no. 1 (2017): 3–14, https://doi.org/10.1680/jcoma.17.00013.

15. K. Zhang, Marwa Alhinai, P. Chermprayong, and Robert Siddall, "SpiderMAV: Perching and Stabilizing Micro Aerial Vehicles with Bio-Inspired Tensile Anchoring Systems," in *Proceedings of the 2017 IEEE/RSJ International Conference on Intelligent Robots and Systems* (Piscataway, NJ: IEEE, 2017), 6849–6854, https://doi.org/10.1109/IROS.2017.8206606.

16. ARCAS was an EU-funded collaborative project that involved eleven research groups and was aimed at development and experimental validation of the first cooperative, free-flying robot system for assembly and structure construction. See European Commission, "Aerial Robotics Cooperative Assembly System," accessed January 25, 2022, https://cordis.europa.eu/project/id/287617.

17. K. Kondak et al., "Aerial Manipulation Robot Composed of an Autonomous Helicopter and a 7 Degrees of Freedom Industrial Manipulator," in *Proceedings of the 2014 IEEE International Conference on Robotics and Automation* (Piscataway, NJ: IEEE, 2014), 2107–2112, https://doi.org/10.1109/ICRA.2014.6907148.

18. Anibal Ollero et al., "The AEROARMS Project: Aerial Robots with Advanced Manipulation Capabilities for Inspection and Maintenance," *IEEE Robotics Automation Magazine* 25, no. 4 (2018): 12–23, https://doi.org/10.1109/MRA.2018.2852789.

CHAPTER 5

1. Bill Gates, "A Robot in Every Home," *Scientific American*, February 1, 2008, https://www.scientificamerican.com/article/a-robot-in-every-home-2008-02/.

2. "Notice of Annual Meeting of Stockholders and iRobot 2021 Proxy Statement," iRobot Corporation, accessed January 25, 2022, https://investor.irobot.com/static-files/0dc4a7a5-cd5a-435e-95db-d3700c3b0bb2.

3. S. E. Galaitsi, Christine Oglivie Hendren, Benjamin Trump, and Igor Linkov, "Sex Robots—A Harbinger for Emerging AI Risk," *Frontiers in Artificial Intelligence* 2 (2019), https://doi.org/10.3389/frai.2019.00027.

4. J. Coopersmith, "Pornography, Videotape and the Internet," in *IEEE Technology and Society Magazine* 19, no. 1 (2000): 27–34, https://doi.org/ 10.1109/44.828561.

5. Nature Editorial, "Let's Talk about Sex Robots," *Nature* 547, no. 138 (2017), https://doi.org/10.1038/547138a.

6. ABC News, "You Can Soon Buy a Sex Robot Equipped with Artificial Intelligence for about $20,000," YouTube video, 7:27, uploaded April 25, 2018, https://youtu.be/-cN8sJz50Ng.

7. Vito Cacucciolo, Jun Shintake, Yu Kuwajima, Shingo Maeda, Dario Floreano, and Herbert Shea, "Stretchable Pumps for Soft Machines," *Nature* 572, no. 7770 (2019): 516–519, https://doi.org/10.1038/s41586-019-1479-6.

8. Jun Shintake, Samuel Rosset, Bryan Schubert, Dario Floreano and Herbert Shea, "Versatile Soft Grippers with Intrinsic Electroadhesion Based on Multifunctional Polymer Actuators," *Advanced Materials* 28, no. 2 (2016): 231–238, https://doi.org/10.1002/adma.201504264.

9. Seyed M. Mirvakili and Ian W. Hunter, "Artificial Muscles: Mechanisms, Applications, and Challenges," *Advanced Materials* 30, no. 6 (2018): 1704407, https://doi.org/10.1002/adma.201704407.

10. Yuki Asano, Kei Okada, and Masayuki Inaba, "Design Principles of a Human Mimetic Humanoid: Humanoid Platform to Study Human Intelligence and Internal Body System," *Science Robotics* 2, no. 13 (2017): eaaq0899, https://doi.org/10.1126/scirobotics.aaq0899.

11. Matt Simon, "A Freaky Humanoid Robot That Sweats as It Does Push-Ups," *Wired*, December 28, 2017, https://www.wired.com/story/a-freaky-humanoid-robot-that-sweats-as-it-does-push-ups.

12. R. I. Damper and Owen Holland, "Exploration and High Adventure: The Legacy of Grey Walter," *Philosophical Transactions of the Royal Society of London. Series A: Mathematical, Physical and Engineering Sciences* 361, no. 1811 (2003): 2085–2121, https://doi.org/10.1098/rsta.2003.1260.

13. Valentino Braitenberg, *Vehicles: Experiments in Synthetic Psychology* (Cambridge, MA: MIT Press, 1986).

14. C, Breazeal, *Designing Sociable Robots* (Cambridge, MA: MIT Press, 2002).

15. Cynthia Breazeal, "Toward Sociable Robots," *Robotics and Autonomous Systems* 42, no. 3 (2003): 167–175, https://doi.org/10.1016/S0921-8890(02)00373-1.

16. Guy Hoffman, "Evaluating Fluency in Human–Robot Collaboration," *IEEE Transactions on Human-Machine Systems* 49, no. 3 (2019): 209–218, https://doi.org/10.1109/THMS.2019.2904558.

17. Kate Darling, *The New Breed: What Our History with Animals Reveals about Our Future with Robots* (New York: Holt, 2021).

18. Jamy Jue Li, Wendy Ju, and Byron Reeves, "Touching a Mechanical Body: Tactile Contact with Body Parts of a Humanoid Robot Is Physiologically Arousing," *Journal of Human-Robot Interaction* 6, no. 3 (2017): 118–130, https://doi.org/10.5898/JHRI.6.3.Li.

19. Kazuyoshi Wada and Takanori Shibata, "Living with Seal Robots—Its Sociopsychological and Physiological Influences on the Elderly at a Care House," *IEEE Transactions on Robotics* 23, no. 5 (2007): 972–980, https://doi.org/10.1109/TRO.2007.906261.

20. K. Wada, T. Shibata, T. Saito and K. Tanie, "Effects of Three Months Robot Assisted Activity to Depression of Elderly People Who Stay at a Health Service Facility for the Aged," in *Proceedings of the SICE 2004 Annual Conference* (2004), 3:2709–2714.

21. Takanori Shibata, "Therapeutic Seal Robot as Biofeedback Medical Device: Qualitative and Quantitative Evaluations of Robot Therapy in Dementia Care," in *Proceedings of the IEEE* 100, no. 8 (2012): 2527–2538, https://doi.org/10.1109/JPROC.2012.2200559.

22. Kazuyoshi Wada and Takanori Shibata, "Social Effects of Robot Therapy in a Care House— Change of Social Network of the Residents for Two Months," in *Proceedings of the 2007 IEEE International Conference on Robotics and Automation* (Piscataway, NJ: IEEE, 2007), 1250–1255, https://doi.org/10.1109/ROBOT.2007.363156.

23. Roland S. Johansson and J. Randall Flanagan, "Coding and Use of Tactile Signals from the Fingertips in Object Manipulation Tasks," *Nature Reviews Neuroscience* 10, no. 5 (2009): 345–359, https://doi.org/10.1038/nrn2621.

24. Gordon Cheng, Florian Bergner, Julio Olvera, Quentin LeBoutet, and Phillip Mittendorfer, "A Comprehensive Realization of Robot Skin: Sensors, Sensing, Control, and Applications," *Proceedings of the IEEE* 107, no. 10 (2019): 2034–2051, https://doi.org/10.1109/JPROC.2019.2933348.

25. Ryan L. Truby, Cosimo Della Santina, and Daniela Rus, "Distributed Proprioception of 3D Configuration in Soft, Sensorized Robots via Deep Learning," *IEEE Robotics and Automation Letters* 5, no. 2 (2020): 3299–3306, https://doi.org/10.1109/LRA.2020.2976320.

26. Alexis E. Block and Katherine J. Kuchenbecker, "Softness, Warmth, and Responsiveness Improve Robot Hugs," *International Journal of Social Robotics* 11, no. 1 (2019): 49–64, https://doi.org/10.1007/s12369-018-0495-2.

27. Shuichi Nishio, Hiroshi Ishiguro, and Norihiro Hagita, "Geminoid: Teleoperated Android of an Existing Person," *Humanoid Robots: New Developments*, June 1, 2007, https://doi.org/10.5772/4876.

28. Masahiro Mori, Karl F. MacDorman, and Norri Kageki, "The Uncanny Valley," *IEEE Robotics Automation Magazine* 19, no. 2 (2012): 98–100, https://doi.org/10.1109/MRA.2012.2192811.

29. Ayse Pinar Saygin, Thierry Chaminade, Hiroshi Ishiguro, Jon Driver, and Chris Frith, "The Thing That Should Not Be: Predictive Coding and the Uncanny Valley in Perceiving Human and Humanoid Robot Actions," *Social Cognitive and Affective Neuroscience* 7, no. 4 (2012): 413–422, https://doi.org/10.1093/scan/nsr025.

30. Girl on the Net, "It's a Sex Robot, but Not as You Know It: Exploring the Frontiers of Erotic Technology," *Guardian*, December 1, 2017, https://www.theguardian.com/science/brain-flapping/2017/dec/01/its-a-sex-robot-but-not-as-you-know-it-exploring-the-frontiers-of-erotic-technology.

31. Charles Q. Choi, "Humans Marrying Robots? A Q&A with David Levy," *Scientific American*, February 19, 2008, https://www.scientificamerican.com/article/humans-marrying-robots/.

32. K. Richardson, "Sex Robot Matters: Slavery, the Prostituted, and the Rights of Machines," in *IEEE Technology and Society Magazine* 35, no. 2 (2016): 46–53, https://doi.org/10.1109 /MTS.2016.2554421.

CHAPTER 6

1.   International Federation of Robotics, "World Robotics Service Robotics," 2020.

2.   Francisco Suárez-Ruiz, Xian Zhou, and Quang-Cuong Pham, "Can Robots Assemble an IKEA Chair?," *Science Robotics* 3, no. 17 (2018), https://doi.org/10.1126/scirobotics .aat6385.

3.   J. Shintake, Vito Cacucciolo, Dario Floreano, and Herbert Shea, "Soft Robotic Grippers," *Advanced Materials*, May 7, 2018, e1707035, https://doi.org/10.1002/adma.201707035.

4.   Raphael Deimel and Oliver Brock, "A Novel Type of Compliant and Underactuated Robotic Hand for Dexterous Grasping," *International Journal of Robotics Research* 35, no. 1–3 (2016): 161–185, https://doi.org/10.1177/0278364915592961.

5.   Rolf Pfeifer and Josh Bongard, *How the Body Shapes the Way We Think* (Cambridge, MA: MIT Press, 2006).

6.   Taekyoung Kim, Sudong Lee, Taehwa Hong, and Gyowook Shin, "Heterogeneous Sensing in a Multifunctional Soft Sensor for Human-Robot Interfaces," *Science Robotics* 5, no. 49 (2020), https://doi.org/10.1126/scirobotics.abc6878.

7.   Wang Wei Lee et al., "A Neuro-Inspired Artificial Peripheral Nervous System for Scalable Electronic Skins," *Science Robotics* 4, no. 32 (2019), https://doi.org/10.1126/scirobotics .aax2198.

8.   Sergey Levine, Peter Pastor, Alex Krizhevsky, and Deirdre Quillen, "Learning Hand-Eye Coordination for Robotic Grasping with Deep Learning and Large-Scale Data Collection," arXiv:1693,02199 (2017), https://drive.google.com/open?id =0B0mFoBMu8f8BaHYzOXZMdzVOalU.

9.   S. Kim, A. Shukla, and A. Billard, "Catching Objects in Flight," *IEEE Transactions on Robotics* 30, no. 5 (2014): 1049–1065, https://doi.org/10.1109/TRO.2014.2316022.

10. K. Kronander and A. Billard, "Learning Compliant Manipulation through Kinesthetic and Tactile Human-Robot Interaction," *IEEE Transactions on Haptics* 7, no. 3 (2014): 367–380, https://doi.org/10.1109/TOH.2013.54.

11. Cecilia Laschi, Matteo Cianchetti, Barbara Mazzolai, Laura Margheri, Maurizio, and Paolo Dario, "Soft Robot Arm Inspired by the Octopus," *Advanced Robotics* 26, no. 7 (2012): 709–727, https://doi.org/10.1163/156855312X626343.

12. Joran W. Booth, Dylan Shah, Jennifer C. Case, Edward L. White, Michelle C. Yuen, Olivier Cyr-Choiniere, and Rebecca Kramer-Bottiglio, "OmniSkins: Robotic Skins That Turn Inanimate Objects into Multifunctional Robots," *Science Robotics* 3, no. 22 (2018): eaat1853, https://doi.org/10.1126/scirobotics.aat1853.

13. D. S. Shah, Michelle C. Yuen, Liana G. Tilton, Ellen J. Yang, and Rebecca Kramer-Bottiglio, "Morphing Robots Using Robotic Skins That Sculpt Clay," *IEEE Robotics and Automation Letters* 4, no. 2 (2019): 2204–11, https://doi.org/10.1109/LRA.2019.2902019.

14. Booth et al., "OmniSkins."

15. J. W. Booth, D. Shah, J. C. Case, E. L. White, M. C. Yuen, O. Cyr-Choiniere, R. Kramer-Bottiglio, "OmniSkins: Robotic Skins That Turn Inanimate Objects into Multifunctional Robots," *Science Robotics* 3, no. 22 (2018): eaat1853.

CHAPTER 7

1. "Bioinspired Robotics #3: Wearables, with Conor Walsh: Robohub," accessed September 17, 2021, https://robohub.org/bioinspired-robotics-3-wearables-with-conor-walsh/.

2. Jesse Dunietz, "Robotic Exoskeleton Adapts While It's Worn," *Scientific American*, July 27, 2017, https://www.scientificamerican.com/article/robotic-exoskeleton-ldquo-evolves -rdquo-while-its-worn/.

3. Sangjun Lee et al., "Autonomous Multi-Joint Soft Exosuit with Augmentation-Power-Based Control Parameter Tuning Reduces Energy Cost of Loaded Walking," *Journal of NeuroEngineering and Rehabilitation* 15 (2018), https://doi.org/10.1186/s12984-018-0410-y.

4. Jinsoo Kim et al., "Reducing the Metabolic Rate of Walking and Running with a Versatile, Portable Exosuit," *Science* 365, no. 6454 (2019): 668–672, https://doi.org/10.1126/science .aav7536.

5. Evelyn J. Park et al., "A Hinge-Free, Non-Restrictive, Lightweight Tethered Exosuit for Knee Extension Assistance during Walking," *IEEE Transactions on Medical Robotics and Bionics* 2, no. 2 (2020): 165–175, https://doi.org/10.1109/TMRB.2020.2989321.

6. F. B. Wagner, J. B. Mignardot, C. G. Le Goff-Mignardot, R. Demesmaeker, S. Komi, M. Capogrosso, et al., "Targeted Neurotechnology Restores Walking in Humans with Spinal Cord Injury," *Nature* 563 (2018): 65–71, https://doi.org/10.1038/s41586-018-0649-2.

7. Jean-Baptiste Mignardot et al., "A Multidirectional Gravity-Assist Algorithm That Enhances Locomotor Control in Patients with Stroke or Spinal Cord Injury," *Science Translational Medicine* 9, no. 399 (2017), https://doi.org/10.1126/scitranslmed.aah3621.

8. David Borton, Silvestro Micera, José del R. Milan, and Grégoire Courtine, "Personalized Neuroprosthetics," *Science Translational Medicine* 5, no. 210 (2013): 210rv2–210rv2, https:// doi.org/10.1126/scitranslmed.3005968.

9. Leigh R. Hochberg et al., "Reach and Grasp by People with Tetraplegia Using a Neurally Controlled Robotic Arm," *Nature* 485, no. 7398 (2012): 372–375, https://doi.org/10.1038 /nature11076.

10. A. Biasiucci et al., "Brain-Actuated Functional Electrical Stimulation Elicits Lasting Arm Motor Recovery after Stroke," *Nature Communications* 9, no. 1 (2018): 2421, https://doi .org/10.1038/s41467-018-04673-z.

11. Grégoire Courtine and Michael V. Sofroniew, "Spinal Cord Repair: Advances in Biology and Technology," *Nature Medicine* 25, no. 6 (2019): 898–908, https://doi.org/10.1038/s41591-019-0475-6.

CHAPTER 8

1. B. Gregory Thompson et al., "Guidelines for the Management of Patients with Unruptured Intracranial Aneurysms," *Stroke* 46, no. 8 (2015): 2368–2400, https://doi.org/10.1161/STR.0000000000000070.

2. Yu Sun and Bradley J. Nelson, "Biological Cell Injection Using an Autonomous Micro-Robotic System—2002," *International Journal of Robotics Research* 21, no. 10–11 (2002), https://journals.sagepub.com/doi/10.1177/0278364902021010833.

3. Felix Beyeler, Adrian Neild, Stefano Oberti, Dominik J. Bell, Yu Sun, Jurg Dual, and Bradley J. Nelson, "Monolithically Fabricated Microgripper with Integrated Force Sensor for Manipulating Microobjects and Biological Cells Aligned in an Ultrasonic Field," *Journal of Microelectromechanical Systems* 16, no. 1 (2007): 7–15, https://doi.org/10.1109/JMEMS.2006.885853.

4. Li Zhang, Jake J. Abbott, Lixin Ding, Bradley E. Kratochvil, Dominik Bell, and Bradley J. Nelson, "Artificial Bacterial Flagella: Fabrication and Magnetic Control," *Applied Physics Letters* 94, no. 6 (2009), https://aip.scitation.org/doi/10.1063/1.3079655.

5. Ben Wang, Kostas Kostarelos, and L. Zhang, "Trends in Micro-/Nanorobotics: Materials Development, Actuation, Localization, and System Integration for Biomedical Applications," *Advanced Materials* 33, no. 4 (2021): 2002047, https://doi.org/10.1002/adma.202002047.

6. Soichiro Tottori, Li Zhang, Famin Qiu, Krzysztof Krawczyk, Alfredo Franco-Obregón, and Bradley J. Nelson, "Magnetic Helical Micromachines: Fabrication, Controlled Swimming, and Cargo Transport," *Advanced Materials* 24, no. 6 (2012), 811–816, https://onlinelibrary.wiley.com/doi/abs/10.1002/adma.201103818.

7. Veronica Iacovacci, Leonardo Ricotti, Edoardo Sinibaldi, Giovanni Signore, Fabio Vistoli, and Arianna Menciassi, "An Intravascular Magnetic Catheter Enables the Retrieval of Nano-agents from the Bloodstream," *Advanced Science* (September 19, 2018), https://onlinelibrary.wiley.com/doi/full/10.1002/advs.201800807.

8. S. Pane, V. Iacovacci, E. Sinibaldi, and A. Menciassi, "Real-Time Imaging and Tracking of Microrobots in Tissues Using Ultrasound Phase Analysis," *Applied Physics Letters* 118, no. 1 (2021): 014102, https://doi.org/10.1063/5.0032969.

9. Leonardo Ricotti et al., "Biohybrid Actuators for Robotics: A Review of Devices Actuated by Living Cells," *Science Robotics* 29, no. 2 (2017), https://robotics.sciencemag.org/content/2/12/eaaq0495.abstract.

10. Yunus Alapan, Berk Yigit, Onur Beker, Ahmet F. Demirörs, and Metin Sitti, "Shape-Encoded Dynamic Assembly of Mobile Micromachines," *Nature Materials* 18, no. 11 (2019): 1244–1251, https://doi.org/10.1038/s41563-019-0407-3.

11. Amirreza Aghakhani, Oncay Yasa, Paul Wrede, and Metin Sitti, "Acoustically Powered Surface-Slipping Mobile Microrobots," *Proceedings of the National Academy of Sciences* 117, no. 7 (2020): 3469–3477, https://doi.org/10.1073/pnas.1920099117.

12. Richard P. Feynman, "There's Plenty of Room at the Bottom," *Engineering and Science* 23, no. 5 (1960): 22–36.

13. "Tiny Batteries Pose Sizeable Risk," National Safety Council, accessed January 25, 2022, https://www.nsc.org/community-safety/safety-topics/child-safety/button-batteries.

14. Shuhei Miyashita , S. Guitron, K. Yoshida, S. Li. D. D. Damian, and D. L. Rus, "Ingestible, Controllable, and Degradable Origami Robot for Patching Stomach Wounds," in *Proceedings of the 2016 IEEE International Conference on Robotics and Automation* (Piscataway, NJ: IEEE, 2016), 909–916, https://doi.org/10.1109/ICRA.2016.7487222.

15. Alexis du Plessis d'Argentré et al., "Programmable Medicine: Autonomous, Ingestible, Deployable Hydrogel Patch and Plug for Stomach Ulcer Therapy," in *Proceedings of the 2018 IEEE International Conference on Robotics and Automation* (Piscataway, NJ: IEEE, 2016), 1511–1518, https://doi.org/10.1109/ICRA.2018.8460615.

CHAPTER 9

1. Oladele A. Ogunseitan, Julie M. Schoenung, Jean-Daniel M. Saphoro, and Andrew A. Shapiro, "The Electronics Revolution: From E-Wonderland to E-Wasteland," *Science* 326, no. 5953 (2009): 670–671, https://doi.org/10.1126/science.1176929.

2. Jonathan Rossiter, Jonathan Winfield, and Ioannis Ieropoulos, "Eating, Drinking, Living, Dying and Decaying Soft Robots," in *Soft Robotics: Trends, Applications and Challenges*, ed. Cecilia Laschi, Jonathan Rossiter, Fumiya Iida, Matteo Cianchetti, and Laura Margheri (Cham: Springer, 2017), 95–101, https://doi.org/10.1007/978-3-319-46460-2_12.

3. Peter Fratzl, "Biomimetic Materials Research: What Can We Really Learn from Nature's Structural Materials?," *Journal of the Royal Society Interface* 4, no. 15 (2007): 637–642, https://doi.org/10.1098/rsif.2007.0218.

4. Ingo Burgert and Peter Fratzl, "Actuation Systems in Plants as Prototypes for Bioinspired Devices," *Philosophical Transactions of the Royal Society A: Mathematical, Physical and Engineering Sciences* 367, no. 1893 (2009): 1541–1557, https://doi.org/10.1098/rsta.2009.0003.

5. Barbara Mazzolai, Virglio Mattoli, Lucia Beccai, and Edoardo Sinibaldi, "Emerging Technologies Inspired by Plants," in *Bioinspired Approaches for Human-Centric Technologies*, ed. Roberto Cingolani (Cham: Springer, 2014), 111–132, https://doi.org/10.1007/978-3-319-04924-3_4.

6. Wenwen Xu et al., "Food-Based Edible and Nutritive Electronics," *Advanced Materials Technologies* 2, no. 11 (2017): 1700181, https://doi.org/10.1002/admt.201700181.

7. Alina S. Sharova et al., "Edible Electronics: The Vision and the Challenge," *Advanced Materials Technologies* 6, no. 2 (2021): 2000757, https://doi.org/10.1002/admt.202000757.

8. Xu et al., "Food-Based Edible and Nutritive Electronics."

9. Sharova et al., "Edible Electronics."

10. Xu Wang et al., "Food-Materials-Based Edible Supercapacitors," *Advanced Materials Technologies* 1, no. 3 (2016): 1600059, https://doi.org/10.1002/admt.201600059.

11. Mostafa Rahimnejad, Arash Adhami, Soleil Darvari, Alireza Zirepour, and Sang-Eun-Oh, "Microbial Fuel Cell as New Technology for Bioelectricity Generation: A Review," *Alexandria Engineering Journal* 54, no. 3 (2015): 745–756, https://doi.org/10.1016/j.aej.2015.03.031.

12. Stuart Wilkinson, "'Gastrobots'—Benefits and Challenges of Microbial Fuel Cells in Food Powered Robot Applications," *Autonomous Robots* 9, no. 2 (2000): 99–111, https://doi.org/10.1023/A:1008984516499.

13. Ian Kelly and Chris Melhuish, "SlugBot: A Robot Predator," in *Advances in Artificial Life*, ed. Jozef Kelemen and Petr Sosík (Berlin: Springer, 2001), 519–528, https://doi.org/10.1007/3-540-44811-X_59.

14. I. Ieropoulos, John Greenman, Chris Melhuish, and Ian Horsfield, "EcoBot-III: A Robot with Guts," in *Proceedings of the 12th International Conference on the Synthesis and Simulation of Living Systems*, University of Southern Denmark, August 19, 2010, https://research-information.bris.ac.uk/en/publications/ecobot-iii-a-robot-with-guts.

15. Hemma Philamore et al., "Row-Bot: An Energetically Autonomous Artificial Water Boatman," in *Proceedings of the 2015 IEEE/RSJ International Conference on Intelligent Robots and Systems* (Piscataway, NJ: IEEE, 2015), 3888–3893, https://doi.org/10.1109/IROS.2015.7353924.

16. Steven Ceron et al., "Popcorn-Driven Robotic Actuators," in *Proceedings of the 2018 IEEE International Conference on Robotics and Automation* (Piscataway, NJ: IEEE, 2018), 4271–4276, https://doi.org/10.1109/ICRA.2018.8461147.

17. Melanie Baumgartner et al., "Resilient Yet Entirely Degradable Gelatin-Based Biogels for Soft Robots and Electronics," *Nature Materials* 19, no. 10 (2020): 1102–1109, https://doi.org/10.1038/s41563-020-0699-3.

18. L. D. Chambers et al., "Biodegradable and Edible Gelatine Actuators for Use as Artificial Muscles," in *Proceedings of the Electroactive Polymer Actuators and Devices Conference* (International Society for Optics and Photonics, 2014), 90560B, https://doi.org/10.1117/12.2045104.

19. Jun Shintake, Harshal Sonar, Egor Piskarev, Jamie Paik, and Dario Floreano, "Soft Pneumatic Gelatin Actuator for Edible Robotics," arXiv:1703.01423 (2017), 6221–6226, https://doi.org/10.1109/IROS.2017.8206525.

20. Josie Hughes and Daniela Rus, "Mechanically Programmable, Degradable Amp; Ingestible Soft Actuators," in *Proceedings of the 2020 3rd IEEE International Conference on Soft Robotics* (Piscataway, NJ: IEEE, 2020), 836–843, https://doi.org/10.1109/RoboSoft48309.2020.9116001.

21. Jie Sun, Weibiao Zhou, Dejian Huang, Jerry Y. H. Fuh, and Geok Soon Hong, "An Overview of 3D Printing Technologies for Food Fabrication," *Food and Bioprocess Technology* 8, no. 8 (2015): 1605–1615, https://doi.org/10.1007/s11947-015-1528-6.

22. Amit Zoran, "Cooking with Computers: The Vision of Digital Gastronomy [Point of View]," *Proceedings of the IEEE* 107, no. 8 (2019): 1467–1473, https://doi.org/10.1109/JPROC.2019.2925262.

23. "ROBOFOOD. The New Science and Technology of Edible Robots and Robotic Food for Humans and Animals," accessed January 25, 2022, https://www.robofood.org.

24. Wen Wang, Lining Yao, Chin-Yi Cheng, Teng Zhang, Daniel Levine, and Hiroshi Ishii, "Transformative Appetite: Shape-Changing Food Transforms from 2D to 3D by Water Interaction through Cooking," in *Proceedings of the 2017 CHI Conference on Human Factors in Computing Systems* (New York: Association for Computing Machinery, 2017), 6123–6132, https://doi.org/10.1145/3025453.3026019.

25. Ye Tao et al., "Morphing Pasta and Beyond," *Science Advances* 7, no. 19 (2021): eabf4098, https://doi.org/10.1126/sciadv.abf4098.

26. Lining Yao, Jifei Ou, Chin-Yi Cheng, Helene Steiner, Wen Wang, Guanyun Wang, Hiroshi Ishii, "BioLogic: Natto Cells as Nanoactuators for Shape Changing Interfaces," in *Proceedings of the 33rd Annual ACM Conference on Human Factors in Computing Systems* (New York: Association for Computing Machinery, 2015), 1–10, https://doi.org/10.1145/2702123.2702611.

27. Viirj Kan, Emma Vargo, Noah Machover, Horoshi Ishii, Serena Pan, Weixuan Chen, and Yasuaki Kakehi, "Organic Primitives: Synthesis and Design of PH-Reactive Materials Using Molecular I/O for Sensing, Actuation, and Interaction," in *Proceedings of the 2017 CHI Conference on Human Factors in Computing Systems* (New York: Association for Computing Machinery, 2017), 989–1000, https://doi.org/10.1145/3025453.3025952.

28. Lisa J. Burton, Nadia Cheng, César Vega, José Andrés, and John W. M. Bush, "Biomimicry and the Culinary Arts," *Bioinspiration and Biomimetics* 8, no. 4 (October 2013): 044003, https://doi.org/10.1088/1748-3182/8/4/044003.

29. Charles Alan Hamilton, Gursel Alici, and Marc in het Panhuis, "3D Printing Vegemite and Marmite: Redefining 'Breadboards,'" *Journal of Food Engineering* 220 (2018): 83–88, https://doi.org/10.1016/j.jfoodeng.2017.01.008.

30. Ayaka Ishii and Itiro Siio, "BubBowl: Display Vessel Using Electrolysis Bubbles in Drinkable Beverages," in *Proceedings of the 32nd Annual ACM Symposium on User Interface Software and Technology* (New York: Association for Computing Machinery, 2019), 619–623, https://doi.org/10.1145/3332165.3347923.

31. AyakaIshii, Manaka Fukushima, Namiki Tanaka, Yasushi Matoba, Kaori Ikematsu, and Itiro Siil, "Electrolysis Bubble Display Based Art Installations," in *Proceedings of the Fifteenth International Conference on Tangible, Embedded, and Embodied Interaction* (New York, USA: Association for Computing Machinery, 2021), 1–9, https://doi.org/10.1145/3430524.3440632.

32. Hiromi Mochiyama, Mitsuhito Ando, Kenji Misu, Teppei Kuroyanagi, "A Study of Potential Social Impacts of Soft Robots with Organic and Edible Bodies by Observation of an Artwork," in *Proceedings of the 2019 IEEE International Conference on Advanced Robotics and Its Social Impacts* (Piscataway, NJ: IEEE, 2019), 208–212, https://doi.org/10.1109/ARSO46408.2019.8948755.

33. Edward A. Shanken, "Life as We Know It and/or Life as It Could Be: Epistemology and the Ontology/Ontogeny of Artificial Life," Leonardo 31, no. 5 (1998): 383–388. https://doi.org/10.2307/1576602.

34. Y. Kuwana, I. Shimoyama, and H. Miura, "Steering Control of a Mobile Robot Using Insect Antennae," in *Proceedings of the 1995 IEEE/RSJ International Conference on Intelligent Robots and Systems: Human Robot Interaction and Cooperative Robots* (Piscataway, NJ: IEEE, 1995), 2:530–535, https://doi.org/10.1109/IROS.1995.526267.

35. R. Holzer and I. Shimoyama, "Locomotion Control of a Bio-Robotic System via Electric Stimulation," in *Proceedings of the 1997 IEEE/RSJ International Conference on Intelligent Robot and Systems. Innovative Robotics for Real-World Applications* (Piscataway, NJ: IEEE, 1997), 3:1514–1519, https://doi.org/10.1109/IROS.1997.656559.

36. Tat Thang Vo Doan, Melvin Y. W. Tan, Xuan Hien Bui, and Hirotaka Sato, "An Ultralightweight and Living Legged Robot," *Soft Robotics* 5, no. 1 (2018): 17–23, https://doi.org/10.1089/soro.2017.0038.

37. Adam W. Feinberg, Alex Feigel, Sergey S. Shevkoplyas, Sean Sheehy, George M. Whitesides, and Kevin Kit Parker, "Muscular Thin Films for Building Actuators and Powering Devices," *Science* 317, no. 5843 (2007): 1366–1370, https://doi.org/10.1126/science.1146885.

38. Sung-Jin Park et al., "Phototactic Guidance of a Tissue-Engineered Soft-Robotic Ray," *Science* 353, no. 6295 (2016): 158–162, https://doi.org/10.1126/science.aaf4292.

39. Caroline Cvetkovic et al., "Three-Dimensionally Printed Biological Machines Powered by Skeletal Muscle," *Proceedings of the National Academy of Sciences* 111 (2014): 201401577, https://doi.org/10.1073/pnas.1401577111.

40. Sam Kriegman, Douglas Blackiston, Michael Levin, and Josh Bongard, "A Scalable Pipeline for Designing Reconfigurable Organisms," *Proceedings of the National Academy of Sciences* 117, no. 4 (2020): 1853–1859, https://doi.org/10.1073/pnas.1910837117.

41. Rolf Pfeifer and Josh Bongard, *How the Body Shapes the Way We Think* (Cambridge, MA: MIT Press, 2006).

42. Stefano Nolfi and Dario Floreano, *Evolutionary Robotics: The Biology, Intelligence, and Technology of Self-Organizing Machines* (Cambridge, MA: MIT Press, 2000).

43. Hod Lipson and Jordan B. Pollack, "Automatic Design and Manufacture of Robotic Lifeforms," *Nature* 406, no. 6799 (2000): 974–978, https://doi.org/10.1038/35023115; Antoine Cully, Jeff Clune, Danesh Tarapore, and Jean-Baptiste Mouret, "Robots That Can Adapt Like Animals," *Nature* 521, no. 7553 (2015): 503–507, https://doi.org/10.1038/nature14422.

44. Douglas Blackiston, Emma Lederer, Sam Kriegman, Simon Garnier, Joshua Bongard, and Michael Levin, "A Cellular Platform for the Development of Synthetic Living Machines," *Science Robotics* 6, no. 52 (2021), https://doi.org/10.1126/scirobotics.abf1571.

45. Sam Kriegman, Douglas Blackiston, Michael Levin, and Josh Bongard, "Kinematic Self-Replication in Reconfigurable Organisms," *Proceedings of the National Academy of Sciences* 118, no. 49 (2021): e2112672118, https://doi.org/10.1073/pnas.2112672118.

CHAPTER 10

1. Jeremy Rifkin, *The End of Work: The Decline of the Global Labor Force and the Dawn of the Post-Market Era* (New York: Putnam, 1995).

2. Erik Brynjolfsson and Andrew McAfee, *Race against the Machine: How the Digital Revolution Is Accelerating Innovation, Driving Productivity, and Irreversibly Transforming Employment and the Economy* (Lexington, MA: Digital Frontier Press, 2011).

3. Martin Ford, *Rise of the Robots: Technology and the Threat of a Jobless Future* (New York: Basic Books, 2015).

4. Daron Acemoglu, David Autor, Jonathon Hazell, and Pascual Restrepo, "AI and Jobs: Evidence from Online Vacancies," NBER working paper 28257 (2020), https://doi.org/10.3386/w28257.

5. "Robots Encroach on Up to 800 Million Jobs around the World," Bloomberg.com, October 20, 2020, https://www.bloomberg.com/news/articles/2020-10-20/robots-encroach-on-up-to-800-million-jobs-around-the-world.

6. Daphne Leprince-Ringuet, "Robots Will Take 50 Million Jobs in the Next Decade. These Are the Skills You'll Need to Stay Employed," ZDNet, accessed April 16, 2021, https://www.zdnet.com/article/robots-will-take-50-million-jobs-in-the-next-decade-these-are-the-skills-youll-need-to-stay-employed/.

7. PriceWaterhouseCooper, "Will Robots Really Steal Our Jobs? An International Analysis of the Potential Long Term Impact of Automation" (2018).

8. "Robots 'to Replace up to 20 Million Factory Jobs' by 2030," *BBC News*, June 26, 2019, https://www.bbc.com/news/business-48760799.

9. Erin Winick, "Every Study We Could Find on What Automation Will Do to Jobs, in One Chart," *MIT Technology Review*, January 25, 2018, https://www.technologyreview.com/2018/01/25/146020/every-study-we-could-find-on-what-automation-will-do-to-jobs-in-one-chart/.

10. Carl Benedikt Frey and Michael A. Osborne, "The Future of Employment: How Susceptible Are Jobs to Computerisation?," *Technological Forecasting and Social Change* 114 (2017): 254–280, https://doi.org/10.1016/j.techfore.2016.08.019.

11. Melanie Arntz, Terry Gregory, and Ulrich Zierahn, "The Risk of Automation for Jobs in OECD Countries: A Comparative Analysis," May 14, 2016, https://doi.org/10.1787/5jlz9h56dvq7-en.

12. McKinsey Global Institute, "Harnessing Automation for a Future That Works," January 12, 2017," https://www.mckinsey.com/featured-insights/digital-disruption/harnessing-automation-for-a-future-that-works.

13. Antonio Paolillo, Nicola Nosengo, Fabrizio Colella, Fabrizio Schiano, William Stewart, Davide Zambrano, Isabelle Chappuis, Rafael Lalive, and Dario Floreano, "Harnessing Automation for a Future That Works," *Science Robotics* (forthcoming, 2022).

14. "Robotics 2020 Multi-Annual Roadmap for Robotics in Europe. SPARC—the Partnership for Robotics in Europe," accessed January 25, 2022, https://www.eu-robotics.net/sparc/upload/Newsroom/Press/2016/files/H2020_Robotics_Multi-Annual_Roadmap_ICT-2017B.pdf.

15. "Tables Created by BLS," accessed April 16, 2021, https://www.bls.gov/oes/tables.htm.

16. Paolillo et al., "Harnessing Automation."

CHAPTER 11

1. Isaac Asimov, *The Complete Robot* (New York: HarperCollins, 2018).

2. Philip Dick, *Do Androids Dream of Electric Sheep?* (New York: Penguin Random House, 1996).

3. International Federation of Robotics, "IFR Industrial Robotics Report" (International Federation of Robotics, 2021).

4. Lukas Schroth, "Drone Market Size 2020–2025 | Drone Industry Insights," accessed June 10, 2021, //droneii.com/the-drone-market-size-2020-2025-5-key-takeaways.

5. International Federation of Robotics, "IFR Service Robotics Report" (Frankfurt, Germany: International Federation of Robotics, 2021).

6. "Colin Angle 1: 14 Failed Business Models," YouTube video, 9:07, uploaded October 23, 2006, by GarrettFrench, https://youtu.be/O4YT6NLOmr4.

7. "(20) Build a Rover, Send It to the Moon, Sell the Movie Rights: 30 Years of iRobot | LinkedIn," accessed July 15, 2021, https://www.linkedin.com/pulse/build-rover-send-moon-sell-movie-rights-30-years-irobot-colin-angle/.

8. International Federation of Robotics, "IFR Industrial Robotics Report" (Frankfurt, Germany: International Federation of Robotics, 2021).

9. WIPO, "WIPO Technology Trends 2019: Artificial Intelligence" (World Intellectual Property Organization., 2019).

EPILOGUE

1. Isaac Asimov, *I, Robot* (Greenwich, CT: Fawcett Publications, 1950).

2. "ROBOETHICS," accessed September 19, 2021, http://www.roboethics.org/sanremo2004/.

3. UK Engineering and Physical Sciences Research Council, "Principles of Robotics," archived in UK National Archives on July 1, 2021, https://webarchive.nationalarchives.gov.uk /ukgwa/20210701125353/https://epsrc.ukri.org/research/ourportfolio/themes/engineering /activities/principlesofrobotics.

4. Alan F. T. Winfield, Christian Blum, and Wenguo Liu, "Towards an Ethical Robot: Internal Models, Consequences and Ethical Action Selection," in *Advances in Autonomous Robotics Systems*, ed. Michael Mistry, Aleš Leonardis, Mark Witkowski, and Chris Melhuish (Cham: Springer, 2014), 85–96, https://doi.org/10.1007/978-3-319-10401-0_8.

5. Alan F. T. Winfield and Marina Jirotka, "The Case for an Ethical Black Box," in *Towards Autonomous Robotic Systems*, ed. Yang Gao, Saber Fallah, Yaochu Jin, and Constantina Lekakou (Cham: Springer, 2017), 262–273, https://doi.org/10.1007/978-3-319-64107-2_21.

6. Joanna J. Bryson, "Robot, All Too Human," *XRDS: Crossroads, The ACM Magazine for Students* 25, no. 3 (2019): 56–59, https://doi.org/10.1145/3313131.

7. Kate Darling, *The New Breed. What Our History with Animals Reveals about Our Future with Robots* (New York: Holt, 2021).

8. Toby Walsh, "Turing's Red Flag," *Communications of the ACM* 59, no. 7 (2016), 34–37, https://cacm.acm.org/magazines/2016/7/204019-turings-red-flag/fulltext.

9. Alan Winfield, "Ethical Standards in Robotics and AI," *Nature Electronics* 2, no. 2 (2019): 46–48, https://doi.org/10.1038/s41928-019-0213-6.

Index

This index mostly covers the nonfiction parts of the book. The fictional sections of chapters 1–9 are indexed only where they provide actual historical or technical information.